Lecture Notes in Physics

Volume 985

The series Lecture Notes in Physics (LNP), founded in 1969, reports new developments in physics research and teaching - quickly and informally, but with a high quality and the explicit aim to summarize and communicate current knowledge in an accessible way. Books published in this series are conceived as bridging material between advanced graduate textbooks and the forefront of research and to serve three purposes:

- to be a compact and modern up-to-date source of reference on a well-defined topic;
- to serve as an accessible introduction to the field to postgraduate students and non-specialist researchers from related areas;
- to be a source of advanced teaching material for specialized seminars, courses and schools.

Both monographs and multi-author volumes will be considered for publication. Edited volumes should however consist of a very limited number of contributions only. Proceedings will not be considered for LNP.

Volumes published in LNP are disseminated both in print and in electronic formats, the electronic archive being available at springerlink.com. The series content is indexed, abstracted and referenced by many abstracting and information services, bibliographic networks, subscription agencies, library networks, and consortia.

Proposals should be sent to a member of the Editorial Board, or directly to the responsible editor at Springer:

Dr Lisa Scalone
Springer Nature
Physics
Tiergartenstrasse 17
69121 Heidelberg, Germany
lisa.scalone@springernature.com

More information about this series at http://www.springer.com/series/5304

Fumihiko Suekane

Quantum Oscillations

A simple principle underlying important aspects of physics

 Springer

Fumihiko Suekane
Research Center for Neutrino Science
Tohoku University
Sendai, Japan

ISSN 0075-8450   ISSN 1616-6361   (electronic)
Lecture Notes in Physics
ISBN 978-3-030-70526-8   ISBN 978-3-030-70527-5   (eBook)
https://doi.org/10.1007/978-3-030-70527-5

This Springer imprint is published by the registered company Springer Nature Switzerland AG
The registered company address is: Gewerbestrasse 11, 6330 Cham, Switzerland

Preface

The field of elementary particle physics is very wide and profound with various subjects and ideas. However, many important physics concepts are rooted in a few principles; quantum mixing and oscillation. The author, who also wrote "Neutrino Oscillations" (Springer, 2015), aims at introducing and explaining the elementary particle physics from the point of view of the quantum oscillations. While most textbooks treat physics subjects vertically, this textbook tries to connect them horizontally with the thread of the quantum mixing and oscillation.

The topics covered in this book start from the spin motion in a magnetic field which introduces the general and complete formula of the two-body mixing and oscillations. Then various subjects, such as the hydrogen HI line, the Weinberg angle (θ_W), the Fermion mass and mixings (Cabibbo angle and CKM matrix), the K^0-$\overline{K^0}$ oscillation and CP violation, the quark structure and mass pattern of hadrons, the neutrino oscillation, and many others, are explained using the formula in omnibus way. The target audience of this book ranges from university students enrolled in science courses to researchers who would like to see particle physics from a new perspective, despite not being their primary field of research. The author hopes this book will help to lower the barriers when trying to understand the broad range of the elementary particle physics.

September 2020

Dr. Fumihiko Suekane
RCNS, Tohoku University
Sendai, Japan

Acknowledgements

This textbook is based on the seminars the author gave at various institutes in Europe as Blaise Pascal Chair (BPC) while staying at APC lab in Paris during 2017–2018. Without that opportunity, this book would not be realized. First, the author would like to thank the Fondation de l'Ecole Normale Supérieure for giving him the honor of BPC and having invited him to stay in France. The author also would like to thank Dr. A. Cabrera, from recommending him for the BPC, to hosting him at the APC lab. Next, the author would like to thank Drs. T. J. Bezerra, M. Grassi and Profs. K. Ishikawa, J. Shirai for giving precious feedbacks by reading the draft of this book. And finally, the author would like to thank Ms. H. Niko and Mr. S. Srinivas, Springer, for helping the author to prepare for the publication of this book.

Contents

1 Basics of the Quantum Oscillation 1
   1.1   Introduction 1
   1.2   State Mixing and Quantum Oscillation.................... 1
         1.2.1   Transition 2
         1.2.2   Mass Eigenstate 3
         1.2.3   Quantum Oscillation.......................... 5

Part I Electromagnetic Interactions

2 Motion of Electron Spin in Magnetic Fields 9
   2.1   Introduction 9
   2.2   Spin-1/2 and Magnetic Field 9
   2.3   The Pauli Equation 9
         2.3.1   Empirical Derivation of the Pauli Equation 9
         2.3.2   Derivation of the Pauli Equation from the Dirac
                  Equation 10
         2.3.3   Physical Meaning of the Pauli Equation............ 11
   2.4   Spin Motion in the Magnetic Field $\vec{B} = (0, 0, B)$;
        The Simplest Case 12
         2.4.1   Energy Eigenstate............................ 12
         2.4.2   Spin Precession 13
   2.5   Spin Motion in Magnetic Field $\vec{B} = (B, 0, 0)$ 14
         2.5.1   Energy Eigenstates 15
         2.5.2   Oscillation and Precession 16
   2.6   Spin Motion in Magnetic Field $\vec{B} = (0, B, 0)$ 17
         2.6.1   Energy Eigenstates 18
         2.6.2   Oscillation and Precession 18
   2.7   Spin Motion in an Arbitrary Magnetic Field:
        $\vec{B} = (B_x, B_y, B_z)$ 19
         2.7.1   Energy Eigenstate............................ 21
         2.7.2   Oscillation and Precession 22

2.8   If Electron Mass is Included   23
Reference   25

3   Hydrogen Hyperfine Splitting and HI Line   27
3.1   Introduction   27
3.2   Spin Structure of the Hydrogen Atom   27
        3.2.1   Oscillation Between $|\Uparrow\Downarrow\rangle$ and $|\Downarrow\Uparrow\rangle$ States   32
3.3   Hydrogen Magnetic Moment Under External Magnetic
        Field   33
3.4   Hydrogen 21 cm Line   37
References   38

4   Anomalous Magnetic Moment   39
4.1   Introduction   39
4.2   Helicity Conservation   39
        4.2.1   Anomalous Magnetic Moment   42
        4.2.2   Measurements   43
References   46

5   Positronium   47
5.1   Introduction   47
5.2 H_K: e^+-e^- Binding State by Electrostatic Potential   49
        5.2.1 H_M: MM-MM Interactions   49
        5.2.2 H_A: Pair Annihilation and Creation   51
        5.2.3 $H_M + H_A$: Both Effects   52
Reference   54

Part II   Higgs Field

6   Weinberg Angle   57
6.1   Introduction   57
6.2   General Formula of Electromagnetic and Weak Interactions . . .   57
        6.2.1   Correspondence to Photon and Z^0   61
6.3   The Origin of the Vector Boson Transitions; Higgs Field   63
6.4   Chirality Dependence of the Weak Interactions   66
6.5   Test of the Electroweak Theory   67
        6.5.1   Measurements of $\sin^2\theta_W$ 68
        6.5.2   Test of the Electroweak Theory   69
References   71

7   Fermion Mass and Chirality Oscillation   73
7.1   Introduction   73
7.2   Chirality   73
7.3   Dirac Equation as Chirality Transition Equation   74
7.4   Decay Effect   76

8   Quark Mass, Cabibbo Angle and CKM Mixing Matrix   79
  8.1   Introduction   79
  8.2   Four-Quark System and Cabibbo Angle, θ_C   79
        8.2.1   Quark Flavor Oscillation   84
        8.2.2   Uncertainty Principle   85
  8.3   Six-Quark System   86
        8.3.1   Measurement of the CKM Matrix Elements   88
        8.3.2   Transition Amplitude $G_{\alpha\beta}$ 92
        8.3.3   Quark Flavor Oscillation   93
  References   94

Part III   Weak Interactions

9 K^0-$\overline{K^0}$ Oscillation and CP Violation   97
  9.1   Introduction   97
  9.2 K^0-$\overline{K^0}$ Oscillation and Prediction of the Charm Quark Mass . . .   97
  9.3   Six-Quark System and CP Violation   104
        9.3.1 $K^0 - \overline{K^0}$ Oscillation of Six-Quark System   106
        9.3.2   Oscillation of K^0 CP Eigenstate   107
  9.4   Discovery of CP Violation and Measurement of α   108
  References   110

Part IV   Strong Interactions

10   Quark Structure of Mesons   113
  10.1   Introduction   113
  10.2 u, d, s-Quark Masses   113
  10.3 π^+-ρ^+ Mass Difference   115
  10.4   Structure of ρ^0, ω and ϕ   120
         10.4.1   Mixing Between $|u\bar{u}\rangle$ and $|d\bar{d}\rangle$ 120
         10.4.2   Mixing Between $|s\bar{s}\rangle$ and $|u\bar{u}\rangle$, $|d\bar{d}\rangle$ Systems   122
         10.4.3   Experimental Confirmation of the Vector Meson
                  Structure   125
  10.5   Structure of π^0, η and η'   126
  10.6   Color Structure of Meson   128
  References   129

11   Quark Structure of Baryons   131
  11.1   Introduction   131
  11.2   Totally Antisymmetric State   131
  11.3 Δ^{++} Baryon   132
         11.3.1 Δ^+ Baryon   134

11.4   Spin-1/2 Baryon   136
        11.4.1   Why Spin-1/2 (uuu) Baryon Does Not Exist?   136
        11.4.2   Quark Structure of Proton.....................   136
        11.4.3 Λ, Σ^0 and Σ^{0*}   137
11.5   Isospin   139

Part V   Unknown Origin

**12   Neutrino Oscillation: Relativistic Oscillation of Three-Flavor
        System**   145
12.1   Introduction   145
12.2   Two-Flavor Oscillation   145
        12.2.1   Neutrino Transition Amplitudes..................   146
        12.2.2   Oscillation   147
        12.2.3   Relativistic Oscillation Probability   148
        12.2.4   Another Way to Derive Relativistic Neutrino
                  Oscillation   151
        12.2.5   A Relation Between Transition Amplitudes
                  and Neutrino Flavor Mass.....................   151
12.3   Three-Flavor Neutrino Oscillation   152
12.4   Measurements of Oscillation Parameters..................   156
        12.4.1 θ_{23} and Δm_{23}^2   157
        12.4.2 θ_{12} and Δm_{12}^2   157
        12.4.3 θ_{13}   158
        12.4.4 δ   158
        12.4.5   Summary of the Measurements   158
References   160

Appendix A: Summary of Parameters and Formulas   161

Appendix B   165

Appendix C   171

Index   173

Basics of the Quantum Oscillation

1

1.1 Introduction

In this textbook, various physics subjects are explained. They may look quite different but actually their phenomena are rooted in the principle of quantum mixing and oscillation. In this chapter, the basic mechanism of the quantum oscillation is explained.

1.2 State Mixing and Quantum Oscillation

In the quantum mechanics, a particle state is expressed by a wave function ψ which satisfies the Schrödinger equation,

$$i\frac{d}{dt}\psi = H\psi \, , \tag{1.1}$$

where H is Hamiltonian which specifies the system.

Let us think of a two-state system consisting of basic states, $|\psi_a\rangle$ and $|\psi_b\rangle$. The readers may regard them as being two spin states, up $|\Uparrow\rangle$ and down $|\Downarrow\rangle$; or a meson system $|K^0\rangle$, $|\overline{K^0}\rangle$; or two neutrino states $|\nu_e\rangle$, $|\nu_\mu\rangle$; or bound quark states in π^0 meson $|u\bar{u}\rangle$, $|d\bar{d}\rangle$.

The general wave function of this two-state system is a superposition of those basic states,

$$|\Psi[t]\rangle = c_a[t]|\psi_a\rangle + c_b[t]|\psi_b\rangle = \begin{pmatrix} c_a \\ c_b \end{pmatrix} \, . \tag{1.2}$$

The time dependence of $\Psi[t]$ comes from the time dependence of the coefficients, $c_a[t]$ and $c_b[t]$, called amplitude. From the normalization condition of $\Psi[t]$, the

© Springer Nature Switzerland AG 2021
F. Suekane, *Quantum Oscillations*, Lecture Notes in Physics 985,
https://doi.org/10.1007/978-3-030-70527-5_1

following relation between $c_a[t]$ and $c_b[t]$ is required,

$$|\Psi[t]|^2 = |c_a[t]|^2 + |c_b[t]|^2 = 1 , \qquad (1.3)$$

where the orthogonal and normalization conditions between $|\psi_a\rangle$ and $|\psi_b\rangle$

$$\langle \psi_a | \psi_b \rangle = \delta_{ab} \qquad (1.4)$$

are used.

1.2.1   Transition

Next, we assume some effect "X" causes a transition between the two basic states,

$$|\psi_a\rangle \Leftrightarrow |\psi_b\rangle . \qquad (1.5)$$

"X" can be a magnetic field which flips the electron spin: $|\Uparrow\rangle \Leftrightarrow |\Downarrow\rangle$; or the second-order weak interactions responsible of $|K^0\rangle \Leftrightarrow |\overline{K^0}\rangle$; or pair annihilation and creation caused by the strong interactions as in $|u\bar{u}\rangle \Leftrightarrow |d\bar{d}\rangle$; or yet unknown origin of the neutrino oscillations $|\nu_e\rangle \Leftrightarrow |\nu_\mu\rangle$; or the Higgs field that generates particle masses and mixings, etc. The transition can be caused by the interaction with external field, like magnetic field or Higgs field, or by internal dynamics like $|K^0\rangle \Leftrightarrow |\overline{K^0}\rangle$ or $|u\bar{u}\rangle \Leftrightarrow |d\bar{d}\rangle$ transitions.

In such the cases, the Hamiltonian has the following structure:

$$H = \begin{pmatrix} 0 & A \\ A & 0 \end{pmatrix} . \qquad (1.6)$$

The Schrödinger equation (1.1) becomes a time development of the wave function as follows:

$$\frac{d}{dt} \begin{pmatrix} c_a \\ c_b \end{pmatrix} = \begin{pmatrix} 0 & -iA \\ -iA & 0 \end{pmatrix} \begin{pmatrix} c_a \\ c_b \end{pmatrix} \Rightarrow \begin{cases} \dot{c}_a = -iAc_b \\ \dot{c}_b = -iAc_a , \end{cases} \qquad (1.7)$$

where A, called transition amplitude, represents the strength of the transition. A is assumed as real number for now. Due to the transition, some part of $|\psi_a\rangle$ becomes $|\psi_b\rangle$ and some part of $|\psi_b\rangle$ becomes $|\psi_a\rangle$ and the general wave function of the system (1.2) changes to

$$\begin{aligned} |\Psi[t+\delta t]\rangle &= c_a[t+\delta t]|\psi_a\rangle + c_b[t+\delta]|\psi_b\rangle \\ &= (c_a[t] - iAc_b[t]\delta t)|\psi_a\rangle + (c_b[t] - iAc_a[t]\delta t)|\psi_b\rangle \end{aligned} \qquad (1.8)$$

after the infinitesimal time δt. Figure 1.1 is a diagram which shows the transition pictorially. We call Eq. (1.7) the "transition equation".

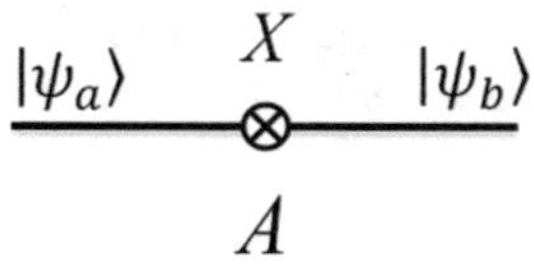

$$|\psi_a\rangle \quad \overset{X}{\underset{\otimes}{\rule{3cm}{0.4pt}}} \quad |\psi_b\rangle$$

$$A$$

Fig. 1.1 Transition diagram that shows the transition between $|\psi_a\rangle$ and $|\psi_b\rangle$. The horizontal line indicates that $|\psi_a\rangle$ changes to $|\psi_b\rangle$ with amplitude $-iA$ per unit time ($c_b \to c_b - iAc_a\delta t$), and vice versa. X represents the cause of the transition. The symbol "$\otimes$" represents the effect of the X onto the states

In order to solve the transition equations (1.7), we add and subtract the two equations and obtain a pair of independent differential equations,

$$\begin{cases} \frac{d}{dt}(c_a + c_b) = -iA(c_a + c_b) \\ \frac{d}{dt}(c_a - c_b) = iA(c_a - c_b) . \end{cases} \tag{1.9}$$

They can easily be solved as

$$\begin{cases} c_a[t] + c_b[t] = c_+ e^{-iAt} \\ c_a[t] - c_b[t] = c_- e^{iAt} , \end{cases} \tag{1.10}$$

where $c_\pm$ are integral constants to be determined by a boundary condition. $c_a[t]$ and $c_b[t]$ can be obtained separately by adding and subtracting the two equations in (1.10) as

$$\begin{cases} c_a[t] = \frac{1}{2}(c_+ e^{-iAt} + c_- e^{iAt}) \\ c_b[t] = \frac{1}{2}(c_+ e^{-iAt} - c_- e^{iAt}) . \end{cases} \tag{1.11}$$

This is the general solution of the transition equation (1.7). By using the general solution, the wave function at an arbitrary time t is expressed as

$$\begin{aligned} |\Psi[t]\rangle &= \frac{1}{2}(c_+ e^{-iAt} + c_- e^{iAt})|\psi_a\rangle + \frac{1}{2}(c_+ e^{-iAt} - c_- e^{iAt})|\psi_b\rangle \\ &= \frac{1}{2}c_+(|\psi_a\rangle + |\psi_b\rangle)e^{-iAt} + \frac{1}{2}c_-(|\psi_a\rangle - |\psi_b\rangle)e^{iAt}. \end{aligned} \tag{1.12}$$

After determining $c_\pm$ we can obtain concrete phenomena behind the general wave function.

1.2.2   Mass Eigenstate

For most of the cases, though not always, we categorize quantum states and particle types by their energies or masses. Therefore, it is important to know the mass or the energy eigenstate of the systems. The wave function of the mass eigenstate with mass m has the following structure:

$$\psi[t] = \text{time independent term} \times e^{-imt}. \tag{1.13}$$

By looking at (1.12), we can immediately recognize the boundary conditions which lead the mass eigenstate of this system as follows:

$$\begin{cases} |\psi_+[t]\rangle = \frac{1}{\sqrt{2}}(|\psi_a\rangle + |\psi_b\rangle)e^{-iAt} \equiv |\psi_+\rangle e^{-iAt}, & \text{for } (c_+, c_-) = (\sqrt{2}, 0) \\ |\psi_-[t]\rangle = \frac{1}{\sqrt{2}}(|\psi_a\rangle - |\psi_b\rangle)e^{+iAt} \equiv |\psi_-\rangle e^{+iAt}, & \text{for } (c_+, c_-) = (0, \sqrt{2}). \end{cases}$$

(1.14)

This means that the mass eigenstate vectors $|\psi_\pm\rangle$ are a mixture of the original vectors $|\psi_a\rangle$ and $|\psi_b\rangle$ as follows:

$$\begin{pmatrix} |\psi_+\rangle \\ |\psi_-\rangle \end{pmatrix} = \frac{1}{\sqrt{2}} \begin{pmatrix} 1 & 1 \\ 1 & -1 \end{pmatrix} \begin{pmatrix} |\psi_a\rangle \\ |\psi_b\rangle. \end{pmatrix}$$

(1.15)

The matrix is called "mixing matrix".

Using these mass eigenstate vectors, $|\psi_\pm\rangle$, the general wave function (1.12) can be expressed concisely as

$$|\Psi[t]\rangle = \frac{1}{\sqrt{2}}c_+ e^{-iAt}|\psi_+\rangle + \frac{1}{\sqrt{2}}c_- e^{iAt}|\psi_-\rangle \equiv C_+[t]|\psi_+\rangle + C_-[t]|\psi_-\rangle. \quad (1.16)$$

This wave function physically means that if we measure the mass of the state $|\Psi\rangle$, we observe the mass $m = A$ with probability $|c_+|^2/2$ and $m = -A$ with probability $|c_-|^2/2$.

It is possible to show that the following relations hold:

$$\begin{cases} \dot{C}_+ = -iAC_+ \\ \dot{C}_- = +iAC_- . \end{cases}$$

(1.17)

If we compare these equations with Eq. (1.7), we notice that Eq. (1.17) describes the transition of a state into itself, namely, $|\psi_+\rangle \Leftrightarrow |\psi_+\rangle$ and $|\psi_-\rangle \Leftrightarrow |\psi_-\rangle$ as shown in Fig. 1.2. We call them "self-transition". On the other hand, the transition shown in Fig. 1.1 changes $|\psi_a\rangle$ to $|\psi_b\rangle$ and $|\psi_b\rangle$ to $|\psi_a\rangle$ and we call it "cross-transition". The transitions shown in Figs. 1.1 and 1.2 are equivalent. Just the base system being treated is different. The mass is actually the transition amplitude of the self-transition.

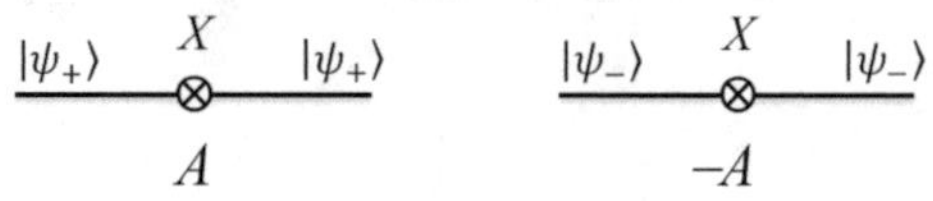

Fig. 1.2 Transition diagram expressed by the mass eigenstate $|\psi_\pm\rangle$. This indicates that the mass is the transition amplitude of the self-transition. These transitions and that in Fig. 1.1 are equivalent

1.2.3   Quantum Oscillation

If the system starts from pure $|\psi_a\rangle$ state, the initial condition is expressed from (1.12) as

$$|\Psi[0]\rangle = |\psi_a\rangle = \frac{1}{2}(c_+ + c_-)|\psi_a\rangle + \frac{1}{2}(c_+ - c_-)|\psi_b\rangle . \tag{1.18}$$

From this relation, the integral constants are determined as

$$c_+ = c_- = 1 . \tag{1.19}$$

By using Eqs. (1.19) and (1.12), the wave function at arbitrary time t is

$$|\Psi[t]\rangle = \cos At|\psi_a\rangle - i \sin At|\psi_b\rangle. \tag{1.20}$$

The probability to find the state $|\psi_b\rangle$ in Ψ at time t is

$$P_{\psi_a \to \psi_b}[t] = |\langle \psi_b|\Psi[t]\rangle|^2 = \sin^2 At . \tag{1.21}$$

At first, the $|\psi_b\rangle$ component was 0 but later it is generated by the transition shown in Fig. 1.1. The probability oscillates in time with angular velocity $\omega = 2A$. This is the simplest example of quantum oscillation. This kind of oscillations plays an important role in various physics phenomena. We will see a variety of the examples in the later sections.

Part I
Electromagnetic Interactions

Motion of Electron Spin in Magnetic Fields

2

2.1 Introduction

The motion of the spin under the influence of a magnetic field is a very good example to introduce quantum mixings and oscillations. In this chapter, the spin motions under various magnetic fields are described. Any two-body system without decay can be understood by analogy with the spin motion of an electron in a magnetic field explained in this chapter.

2.2 Spin-1/2 and Magnetic Field

The motion of the spin under magnetic field can be treated using the Pauli equation. First, we will form the Pauli equation by quantizing the equation of the classical electrodynamics and solve it. By specifying the initial condition, we will then see various oscillation phenomena. The general solution obtained and the phenomena observed here are the core of two-body quantum mechanics and can be directly applied to other physics cases by just replacing the parameters.

2.3 The Pauli Equation

2.3.1 Empirical Derivation of the Pauli Equation

The Pauli equation was constructed empirically by W. Pauli to explain the electromagnetic interaction associated with the electron spin. Later, it turned out that both the Pauli equation and the Bohr magneton can be obtained from the low energy approximation of the Dirac equation for an electron with the electromagnetic interactions.

© Springer Nature Switzerland AG 2021
F. Suekane, *Quantum Oscillations*, Lecture Notes in Physics 985,
https://doi.org/10.1007/978-3-030-70527-5_2

From classical electrodynamics, the energy shift of the electron with magnetic moment $\vec{\mu}_e$ in a magnetic field $\vec{B}$ is expressed as

$$E_B = -\vec{\mu}_e \cdot \vec{B}. \tag{2.1}$$

In order to quantize the classical equation (2.1), we replace the spin vector and energy as follows,[1]

$$\vec{\mu}_e \rightarrow -\mu_B \vec{\sigma}, \qquad E_B \rightarrow H_B = \mu_B(\vec{\sigma} \cdot \vec{B}), \tag{2.2}$$

where μ_B is the Bohr magneton,

$$\mu_B = \frac{e}{2m_e} = 5.788 \times 10^{-5}[\text{eV/T}] \tag{2.3}$$

and $\vec{\sigma}$ is the Pauli sigma vector (A.7).

The equation of the spin motion is then

$$i\frac{d}{dt}\psi = H_B\psi = \mu_B(\vec{\sigma} \cdot \vec{B})\psi = \mu_B \begin{pmatrix} B_z & B_- \\ B_+ & -B_z \end{pmatrix}\psi, \tag{2.4}$$

where ψ is the two-component spinor,

$$\psi[t] = \begin{pmatrix} \alpha[t] \\ \beta[t] \end{pmatrix}. \tag{2.5}$$

Equation (2.4) is called Pauli equation.

2.3.2   Derivation of the Pauli Equation from the Dirac Equation

The Dirac equation of the electron with electromagnetic interaction is

$$(\gamma^\mu(i\partial_\mu + eA_\mu) - m_e)\psi = 0, \tag{2.6}$$

where A_μ is the electromagnetic field. ψ is a four-component spinor; $\psi = \begin{pmatrix} u \\ v \end{pmatrix}$, with u and v being two-component spinors. For low energy approximation, the Dirac equation (2.6) becomes,[2]

$$\left\{ \frac{1}{2m_e}(-i\vec{\nabla} + e\vec{A})^2 + \frac{e}{2m_e}(\vec{\sigma} \cdot (\vec{\nabla} \times \vec{A})) - eA_0 \right\} u = (E - m_e)u. \tag{2.7}$$

---

[1]Like other quantization procedures, there is no a priori theoretical motivation to carry out this operation. However, it is justified by the empirical evidence that all the results obtained from the equations agree with our observation.
[2]For example, Eq. (5.31) of [1].

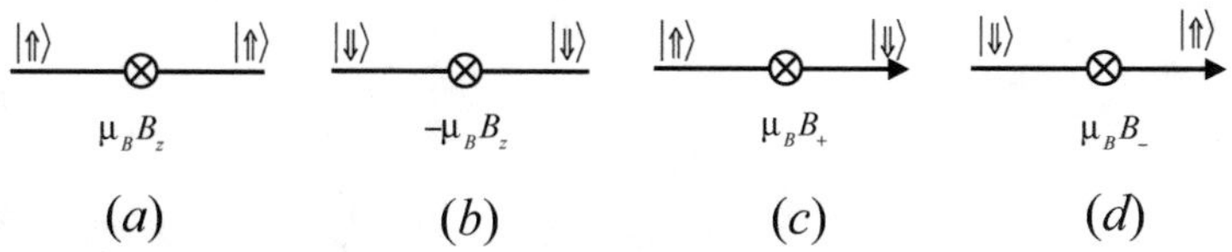

Fig. 2.1 Spin transition amplitudes in a magnetic field $\vec{B}$. B_z component keeps the z-component of the spin direction but B_x and B_y components flip the direction. **a** and **b** are self-transition and **c** and **d** are cross-transition

Remembering that $\vec{B} = (\vec{\nabla} \times \vec{A})$, the second term of Eq. (2.7) can be understood as

$$H_P = \mu_B(\vec{\sigma} \cdot \vec{B}), \quad \mu_B = \frac{e}{2m_e} \ . \tag{2.8}$$

Therefore, the spin and the Bohr magneton can be derived from the relativistic quantum mechanics.

2.3.3 Physical Meaning of the Pauli Equation

The time development of a wave function is, using the infinitesimal δt,

$$\psi[t + \delta t] \sim \psi[t] + \frac{d}{dt}\psi[t]\delta t \ . \tag{2.9}$$

By replacing the time derivative term with Eqs. (2.4), (2.9) becomes

$$\begin{pmatrix} \alpha[t + \delta t] \\ \beta[t + \delta t] \end{pmatrix} = \begin{pmatrix} \alpha[t] \\ \beta[t] \end{pmatrix} - i\mu_B \begin{pmatrix} B_z & B_- \\ B_+ & -B_z \end{pmatrix} \begin{pmatrix} \alpha[t] \\ \beta[t] \end{pmatrix} \delta t \ . \tag{2.10}$$

This is equivalent to the following equations:

$$\begin{cases} \alpha[t + \delta t] = (1 - i\omega_z\delta t)\alpha[t] - i\omega_-\beta[t]\delta t \rightarrow e^{-i\omega_z\delta t}\alpha[t] - i\omega_-\beta[t]\delta t \\ \beta[t + \delta t] = (1 + i\omega_z\delta t)\beta[t] - i\omega_+\alpha[t]\delta t \rightarrow e^{+i\omega_z\delta t}\beta[t] - i\omega_+\alpha[t]\delta t, \end{cases} \tag{2.11}$$

where $\omega_z = \mu_B B_z$ and $\omega_\pm = \mu_B B_\pm$. These equations mean that the amplitude of $|\Uparrow\rangle$, that is α, increases per unit time, as much as the product of the amplitude of $|\Downarrow\rangle$, that is β, and the factor $-i\omega_-$, as well as $-i\omega_z$ times its own amplitude, α. Figure 2.1 shows these spin transitions pictorially. We call the spin-flipping transition ((c), (d)) as *cross-transition* and spin-keeping transition ((a), (b)) as *self-transition*. For cross-transitions, an arrow is necessary since the transition amplitude generally depends on the direction of the change but for self-transition, there is no arrow because the initial and final state are the same.

Fig. 2.2 Spin in the magnetic field $\vec{B} = (0, 0, B)$

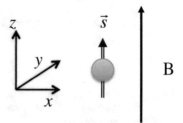

2.4   Spin Motion in the Magnetic Field $\vec{B} = (0, 0, B)$; The Simplest Case

If the magnetic field points to the $+z$ direction as shown in Fig. 2.2, the magnetic field vector is $\vec{B} = (0, 0, B)$ and the Pauli equation becomes

$$i\frac{d}{dt}\begin{pmatrix} \alpha \\ \beta \end{pmatrix} = \mu_B B \sigma_z \begin{pmatrix} \alpha \\ \beta \end{pmatrix} = \omega_B \begin{pmatrix} 1 & 0 \\ 0 & -1 \end{pmatrix} \begin{pmatrix} \alpha \\ \beta \end{pmatrix} = \omega_B \begin{pmatrix} \alpha \\ -\beta \end{pmatrix}; \quad \omega_B \equiv \mu_B B.$$
(2.12)

This equation can directly be solved as

$$\alpha[t] = \alpha[0]e^{-i\omega_B t}, \quad \beta[t] = \beta[0]e^{i\omega_B t}.$$
(2.13)

The general spin wave function is, therefore,

$$\psi[t] = \begin{pmatrix} \alpha[0]e^{-i\omega_B t} \\ \beta[0]e^{i\omega_B t} \end{pmatrix}.$$
(2.14)

2.4.1   Energy Eigenstate

A system is energy eigenstate[3] with energy E if its wave function has the form Eq. (1.13). If the initial condition is

$$\psi[0] = \begin{pmatrix} \alpha[0] \\ \beta[0] \end{pmatrix} = \begin{pmatrix} 1 \\ 0 \end{pmatrix},$$
(2.15)

the wave function at a time t is

$$\psi_{+z}[t] = \begin{pmatrix} 1 \\ 0 \end{pmatrix} e^{-i\omega_B t}.$$
(2.16)

This is the energy eigenstate with energy $E = \omega_B$.

The spin direction can be defined as the direction of the magnetic field which gives the energy $E = \omega_B$. Since the state expressed by Eq. (2.16) has energy ω_B in

---

[3]Energy eigenstate and mass eigenstate are the same.

the magnetic field which points to $+z$ direction, it is the wave function of the state whose spin points to $+z$ direction.[4] The base state vectors for spin pointing to the $\pm z$ direction, $|\Uparrow\rangle$, $|\Downarrow\rangle$ are defined as

$$| \Uparrow \rangle \equiv \begin{pmatrix} 1 \\ 0 \end{pmatrix}, \quad | \Downarrow \rangle \equiv \begin{pmatrix} 0 \\ 1 \end{pmatrix}. \tag{2.17}$$

Using this expression, the wave function (2.5) can be written as

$$\psi[t] = \alpha[t]| \Uparrow \rangle + \beta[t]| \Downarrow \rangle. \tag{2.18}$$

2.4.2 Spin Precession

The wave function of the spin which points to the polar angle (θ, ϕ) is expressed as[5]

$$\hat{s}[\theta, \phi] = \begin{pmatrix} e^{-i(\phi/2)} \cos[\theta/2] \\ e^{i(\phi/2)} \sin[\theta/2] \end{pmatrix}. \tag{2.19}$$

Therefore, if the spin points to the direction (θ_0, ϕ_0) at time $t = 0$, the initial condition is

$$\psi[0] = \begin{pmatrix} \alpha[0] \\ \beta[0] \end{pmatrix} = \hat{s}[\theta_0, \phi_0] = \begin{pmatrix} e^{-i(\phi_0/2)} \cos[\theta_0/2] \\ e^{i(\phi_0/2)} \sin[\theta_0/2]. \end{pmatrix} \tag{2.20}$$

The spin wave function at the arbitrary time t is then,

$$\psi[t] = \begin{pmatrix} \alpha[0]e^{-i\omega_B t} \\ \beta[0]e^{i\omega_B t} \end{pmatrix} = \begin{pmatrix} e^{-i(\phi_0/2 + \omega_B t)} \cos[\theta_0/2] \\ e^{i(\phi_0/2 + \omega_B t)} \sin[\theta_0/2] \end{pmatrix} = \hat{s}[\theta_0, \phi_0 + 2\omega_B t]. \tag{2.21}$$

This means that the spin direction rotates with an angular velocity $2\omega_B$ around the z-axis with a polar angle θ_0. This is called the spin precession shown in Fig. 2.3.

Fig. 2.3 Spin precession under the magnetic field $\vec{B} = (0, 0, B)$. At $t = 0$, the spin pointed to the direction (θ_0, ϕ_0). The spin direction rotates with angular velocity $2\omega_B$ around the z-axis with polar angle θ_0

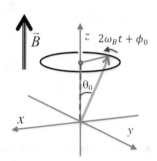

---

[4]There is an ambiguity of overall imaginary phase factor $e^{i\delta}$ which has to be determined by a boundary condition. This point will be discussed later. For the time being, we safely ignore this imaginary phase.

[5]This form is derived in Eq. (2.74).

2.5   Spin Motion in Magnetic Field $\vec{B} = (B, 0, 0)$

If the magnetic field is applied to $+x$ direction as shown in Fig. 2.4, the Pauli equation becomes

$$\frac{d}{dt}\begin{pmatrix} \alpha \\ \beta \end{pmatrix} = -i\mu_B B \sigma_x \begin{pmatrix} \alpha \\ \beta \end{pmatrix} = -i\omega_B \begin{pmatrix} 0 & 1 \\ 1 & 0 \end{pmatrix} \begin{pmatrix} \alpha \\ \beta \end{pmatrix} = -i\omega_B \begin{pmatrix} \beta \\ \alpha \end{pmatrix} . \tag{2.22}$$

In this case the spin transition amplitudes are shown as in Fig. 2.5. Contrary to the previous case ($\vec{B} = (0, 0, B)$), the spin direction changes with amplitude $\mu_B B$, while there is no self-transitions. The corresponding simultaneous differential equations are

$$\begin{cases} \dot{\alpha} = -i\omega_B \beta \\ \dot{\beta} = -i\omega_B \alpha . \end{cases} \tag{2.23}$$

By adding and subtracting the two equations, we obtain the following two independent differential equations:

$$\begin{cases} \dot{\alpha} + \dot{\beta} = -i\omega_B(\alpha + \beta) \\ \dot{\alpha} - \dot{\beta} = +i\omega_B(\alpha - \beta) . \end{cases} \tag{2.24}$$

They can be solved as

$$\begin{cases} \alpha[t] + \beta[t] = (\alpha[0] + \beta[0])e^{-i\omega_B t} \\ \alpha[t] - \beta[t] = (\alpha[0] - \beta[0])e^{+i\omega_B t} . \end{cases} \tag{2.25}$$

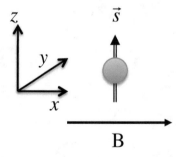

Fig. 2.4 Spin in the magnetic field $\vec{B} = (B, 0, 0)$

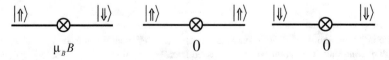

Fig. 2.5 Spin transition amplitude in the magnetic field $\vec{B} = (B, 0, 0)$

By adding and subtracting the equations in Eq. (2.25) again, we can obtain $\alpha[t]$ and $\beta[t]$ separately,

$$\begin{pmatrix} \alpha[t] \\ \beta[t] \end{pmatrix} = \begin{pmatrix} \cos \omega_B t & -i \sin \omega_B t \\ -i \sin \omega_B t & \cos \omega_B t \end{pmatrix} \begin{pmatrix} \alpha[0] \\ \beta[0] \end{pmatrix} . \tag{2.26}$$

The spin wave function is then

$$\psi[t] = \alpha[t] | \Uparrow \rangle + \beta[t] | \Downarrow \rangle$$
$$= (\alpha[0] \cos \omega_B t - i\beta[0] \sin \omega_B t) | \Uparrow \rangle + (\beta[0] \cos \omega_B t - i\alpha[0] \sin \omega_B t) | \Downarrow \rangle . \tag{2.27}$$

2.5.1 Energy Eigenstates

By using the relations

$$\cos \omega t = \frac{e^{i\omega t} + e^{-i\omega t}}{2} \quad \text{and} \quad \sin \omega t = \frac{e^{i\omega t} - e^{-i\omega t}}{2i}, \tag{2.28}$$

Equation (2.27) can be written by grouping with $e^{\pm i\omega_B t}$,

$$\psi[t] = \frac{\alpha[0] + \beta[0]}{\sqrt{2}} \frac{| \Uparrow \rangle + | \Downarrow \rangle}{\sqrt{2}} e^{-i\omega_B t} + \frac{\alpha[0] - \beta[0]}{\sqrt{2}} \frac{| \Uparrow \rangle - | \Downarrow \rangle}{\sqrt{2}} e^{i\omega_B t} . \tag{2.29}$$

From the definition of energy eigenstate Eq. (1.13), if $\alpha[0] = \beta[0] = 1/\sqrt{2}$, this state is the energy eigenstate with energy $E = \omega_B$,

$$\psi_{+x}[t] = \frac{| \Uparrow \rangle + | \Downarrow \rangle}{\sqrt{2}} e^{-i\omega_B t} . \tag{2.30}$$

On the other hand, if $\alpha[0] = -\beta[0] = 1/\sqrt{2}$, the state Eq. (2.29) becomes the energy eigenstate with energy $E = -\omega_B$,

$$\psi_{-x}[t] = \frac{| \Uparrow \rangle - | \Downarrow \rangle}{\sqrt{2}} e^{i\omega_B t} . \tag{2.31}$$

Because the energy of ψ_{+x} state is $+\mu_B B$, the spin directions of this state is pointing to the $+x$ axis. If we write $| \Rightarrow \rangle$ as the basis spin vector which points to the $+x$ direction and $| \Leftarrow \rangle$, $-x$ direction, they are expressed as

$$| \Rightarrow \rangle = \frac{1}{\sqrt{2}} (| \Uparrow \rangle + | \Downarrow \rangle), \quad | \Leftarrow \rangle = \frac{1}{\sqrt{2}} (| \Uparrow \rangle - | \Downarrow \rangle).) \tag{2.32}$$

This relation indicates that the spin polarized to $\pm x$-direction are superposition of $| \Uparrow \rangle$ and $| \Downarrow \rangle$. If we measure $| \Uparrow \rangle$ and $| \Downarrow \rangle$ component in the state $| \Rightarrow \rangle$, we will find $| \Uparrow \rangle$ with 50% probability and $| \Downarrow \rangle$ with 50% probability.

Contrarily, from Eq. (2.32),

$$| \Uparrow \rangle = \frac{1}{\sqrt{2}}(| \Rightarrow \rangle + | \Leftarrow \rangle), \qquad | \Downarrow \rangle = \frac{1}{\sqrt{2}}(| \Rightarrow \rangle - | \Leftarrow \rangle). \qquad (2.33)$$

This means the spin-up state is a superposition of spin $+x$ and spin $-x$ states.

2.5.2 Oscillation and Precession

If the spin points to $+z$ direction ($| \Uparrow \rangle$) at $t = 0$, the initial condition is, from Eq. (2.27);

$$\psi[0] = \alpha[0]| \Uparrow \rangle + \beta[0]| \Downarrow \rangle = | \Uparrow \rangle . \qquad (2.34)$$

$\alpha[0] = 1$ and $\beta[0] = 0$ are derived from this requirement. The wave function at time t becomes

$$\psi[t] = \cos \omega_B t | \Uparrow \rangle - i \sin \omega_B t | \Downarrow \rangle . \qquad (2.35)$$

This means that due to the magnetic field, the spin flip occurs and the probability that the spin is $| \Downarrow \rangle$ state at time t is

$$P_{\Uparrow \rightarrow \Downarrow}[t] = |\langle \Downarrow |\psi[t]\rangle|^2 = \sin^2 \omega_B t . \qquad (2.36)$$

Therefore, the probability to observe $| \Downarrow \rangle$ oscillates as a function of time.

The spin wave function (2.35) can be written as

$$\psi[t] = e^{i(3/4)\pi} \begin{pmatrix} e^{-i(3/4)\pi} \cos \omega_B t \\ e^{i(3/4)\pi} \sin \omega_B t \end{pmatrix} . \qquad (2.37)$$

The overall phase does not give physical effect. Comparing with Eq. (2.19), the spin direction at time t is

$$(\theta[t], \phi[t]) = (2\omega_B t, (3/2)\pi) . \qquad (2.38)$$

The relation (2.38) means the spin direction is precessing in the yz-plane with angular velocity $2\omega_B$ as shown in Fig. 2.6.

Fig. 2.6 Spin precession
under $\vec{B} = (B, 0, 0)$

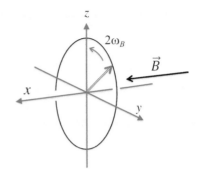

2.6   Spin Motion in Magnetic Field $\vec{B} = (0, B, 0)$

If the magnetic field points to $+y$ direction, the Pauli equation in the magnetic field
is expressed as

$$\frac{d}{dt}\begin{pmatrix} \alpha \\ \beta \end{pmatrix} = -i\mu_B B \sigma_y \begin{pmatrix} \alpha \\ \beta \end{pmatrix} = -i\omega_B \begin{pmatrix} 0 & -i \\ i & 0 \end{pmatrix}\begin{pmatrix} \alpha \\ \beta \end{pmatrix} = -i\omega_B \begin{pmatrix} -i\beta \\ i\alpha \end{pmatrix}. \quad (2.39)$$

The corresponding transition amplitudes are shown in Fig. 2.7. In this case, there is
no self-transitions and the cross-transition amplitudes are pure imaginary number.
The signs of the cross-transition amplitude are opposite depending on the direction
of the transition ($\Uparrow \to \Downarrow$ or $\Downarrow \to \Uparrow$). The simultaneous differential equations from the
Pauli equation are

$$\begin{cases} \dot{\alpha} = -\omega_B \beta \\ \dot{\beta} = +\omega_B \alpha . \end{cases} \quad (2.40)$$

These equations can be solved by the same add-subtract method used in the previous
section,

$$\begin{cases} \dot{\alpha} + i\dot{\beta} = +i\omega_B(\alpha + i\beta) \\ \dot{\alpha} - i\dot{\beta} = -i\omega_B(\alpha - i\beta) . \end{cases} \quad (2.41)$$

This can easily be solved as

$$\begin{cases} \alpha[t] + i\beta[t] = (\alpha[0] + i\beta[0])e^{i\omega_B Bt} \\ \alpha[t] - i\beta[t] = (\alpha[0] - i\beta[0])e^{-i\omega_B Bt} . \end{cases} \quad (2.42)$$

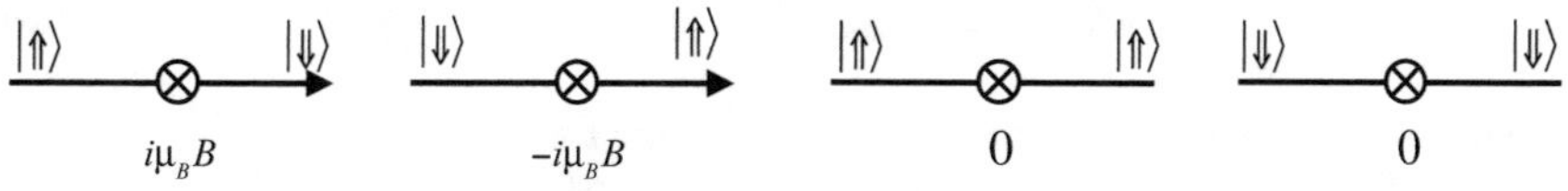

$$|\Uparrow\rangle \underset{i\mu_B B}{\otimes} |\Downarrow\rangle \qquad |\Downarrow\rangle \underset{-i\mu_B B}{\otimes} |\Uparrow\rangle \qquad |\Uparrow\rangle \underset{0}{\otimes} |\Uparrow\rangle \qquad |\Downarrow\rangle \underset{0}{\otimes} |\Downarrow\rangle$$

Fig. 2.7 Spin transition amplitude in the magnetic field $\vec{B} = (0, B, 0)$

By adding and subtracting the equations again, we can obtain $\alpha[t]$ and $\beta[t]$ separately,

$$\begin{pmatrix} \alpha[t] \\ \beta[t] \end{pmatrix} = \begin{pmatrix} \cos \omega_B t & -\sin \omega_B t \\ \sin \omega_B t & \cos \omega_B t \end{pmatrix} \begin{pmatrix} \alpha[0] \\ \beta[0] \end{pmatrix} . \tag{2.43}$$

The general wave function is then

$$\psi[t] = (\alpha[0] \cos \omega_B t - \beta[0] \sin \omega_B t)| \Uparrow \rangle + (\beta[0] \cos \omega_B t + \alpha[0] \sin \omega_B t)| \Downarrow \rangle . \tag{2.44}$$

2.6.1 Energy Eigenstates

The wave function (2.44) can be expressed grouping by the $e^{\pm i \omega_B t}$ terms as

$$\psi[t] = \frac{\alpha[0] - i\beta[0]}{\sqrt{2}} \frac{| \Uparrow \rangle + i| \Downarrow \rangle}{\sqrt{2}} e^{-i\omega_B t} + \frac{\alpha[0] + i\beta[0]}{\sqrt{2}} \frac{| \Uparrow \rangle - i| \Downarrow \rangle}{\sqrt{2}} e^{i\omega_B t} . \tag{2.45}$$

The energy eigenstates can be obtained with the same procedure in the previous subsection as

$$\psi_{+y}[t] = \frac{| \Uparrow \rangle + i| \Downarrow \rangle}{\sqrt{2}} e^{-i\omega_B t}, \quad \psi_{-y}[t] = \frac{| \Uparrow \rangle - i| \Downarrow \rangle}{\sqrt{2}} e^{+i\omega_B t}. \tag{2.46}$$

The spin direction of ψ_{+y} state is along the $+y$ axis. If we write $|\otimes\rangle$ as the basis spin state which points to the $+y$ direction and $|\odot\rangle$, to $-y$ direction, they can be expressed by a superposition of $| \Uparrow \rangle$ and $| \Downarrow \rangle$.

$$|\otimes\rangle = \frac{| \Uparrow \rangle + i| \Downarrow \rangle}{\sqrt{2}} = \frac{1}{\sqrt{2}} \begin{pmatrix} 1 \\ i \end{pmatrix}, \quad |\odot\rangle = \frac{| \Uparrow \rangle - i| \Downarrow \rangle}{\sqrt{2}} = \frac{1}{\sqrt{2}} \begin{pmatrix} 1 \\ -i, \end{pmatrix} \tag{2.47}$$

where the relative coefficients of $| \Uparrow \rangle$ and $| \Downarrow \rangle$ is an imaginary number.

2.6.2 Oscillation and Precession

If the initial spin state is $\psi[0] = | \Uparrow \rangle$, the integral constants in Eq. (2.44) are determined as $(\alpha[0], \beta[0]) = (1, 0)$ and the wave function becomes

$$\psi[t] = \cos \omega_B t | \Uparrow \rangle + \sin \omega_B t | \Downarrow \rangle . \tag{2.48}$$

The probability that the spin is $| \Downarrow \rangle$ state at time t is

$$P_{\Uparrow \to \Downarrow}[t] = |\langle \Downarrow | \psi[t] \rangle|^2 = \sin^2 \omega_B t. \tag{2.49}$$

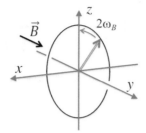

Fig. 2.8 Spin precession under magnetic field $\vec{B} = (0, B, 0)$

The spin wave function (2.48) can be written as

$$\psi[t] = \begin{pmatrix} \cos \omega_B t \\ \sin \omega_B t \end{pmatrix} . \tag{2.50}$$

Comparing with Eq. (2.19), the spin direction at time t is

$$(\theta[t], \phi[t]) = (2\omega_B t, 0) . \tag{2.51}$$

This means that the spin is rotating within the xz-plane with angular velocity $2\omega_B$ as shown in Fig. 2.8. Although the oscillation probabilities Eqs. (2.36) and (2.49) are the same, the physical phenomena in the two cases are different. For Eq. (2.36) case, the spin is precessing within yz-plane and for Eq. (2.49) case, the spin is precessing within zx-plane. The probabilities Eqs. (2.36) and (2.49) are the same because the z-component of the spins at time t are the same for both cases.

2.7 Spin Motion in an Arbitrary Magnetic Field: $\vec{B} = (B_x, B_y, B_z)$

The arbitrary magnetic field vector can be expressed as

$$\vec{B} = (B_x, B_y, B_z) = B (\sin \theta_B \cos \phi_B, \ \sin \theta_B \sin \phi_B, \ \cos \theta_B) . \tag{2.52}$$

In this case, the Pauli equation is written as

$$\frac{d}{dt} \begin{pmatrix} \alpha \\ \beta \end{pmatrix} = -i\mu_B \begin{pmatrix} B_z & B_- \\ B_+ & -B_z \end{pmatrix} \begin{pmatrix} \alpha \\ \beta \end{pmatrix} = -i\omega_B \begin{pmatrix} \cos \theta_B & e^{-i\phi_B} \sin \theta_B \\ e^{i\phi_B} \sin \theta_B & -\cos \theta_B \end{pmatrix} \begin{pmatrix} \alpha \\ \beta \end{pmatrix} . \tag{2.53}$$

The corresponding transition amplitudes are shown Fig. 2.9.

The add-subtract method can be used also here to solve Eq. (2.53). First, we add α and β with arbitrary relative weight λ and obtain

$$\frac{d}{dt}(\alpha + \lambda\beta) = -i\omega_B(\cos \theta_B + \lambda e^{i\phi_B} \sin \theta_B) \left(\alpha + \left(\frac{e^{-i\phi_B} \sin \theta_B - \lambda \cos \theta_B}{\cos \theta_B + \lambda e^{i\phi_B} \sin \theta_B} \right) \beta \right) . \tag{2.54}$$

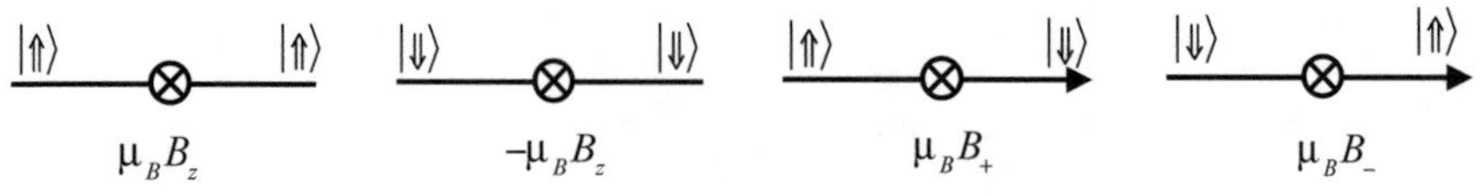

Fig. 2.9 Spin transition amplitude in the arbitoraly magnetic field

If λ satisfies the condition

$$\frac{e^{-i\phi_B} \sin\theta_B - \lambda\cos\theta_B}{\cos\theta_B + \lambda e^{i\phi_B}\sin\theta_B} = \lambda, \tag{2.55}$$

Eq. (2.54) becomes

$$\frac{d}{dt}(\alpha + \lambda\beta) = -i\omega_B(\cos\theta_B + \lambda e^{i\phi_B}\sin\theta_B)(\alpha + \lambda\beta) . \tag{2.56}$$

The general solution is

$$\alpha[t] + \lambda\beta[t] = (\alpha[0] + \lambda\beta[0])\exp[-i\omega_B(\cos\theta_B + \lambda e^{i\phi_B}\sin\theta_B)t] . \tag{2.57}$$

There are two λ's that satisfy Eq. (2.55);

$$\begin{cases} \lambda_+ = \ e^{-i\phi_B}\tan[\theta_B/2] \\ \lambda_- = -e^{-i\phi_B}\cot[\theta_B/2] . \end{cases} \tag{2.58}$$

Since

$$\cos\theta_B + \lambda_\pm e^{i\phi_B}\sin\theta_B = \pm 1, \tag{2.59}$$

Equation (2.57) becomes

$$\alpha[t] + \lambda_\pm\beta[t] = G_\pm\exp[\mp i\omega_B t] , \tag{2.60}$$

where $G_\pm$ are

$$\begin{cases} G_+ = \alpha[0] + e^{-i\phi_B}\tan[\theta_B/2]\beta[0] \\ G_- = \alpha[0] - e^{-i\phi_B}\cot[\theta_B/2]\beta[0] . \end{cases} \tag{2.61}$$

By adding and subtracting Eqs. (2.60), we can obtain $\alpha[t]$ and $\beta[t]$ separately as

$$\begin{cases} \alpha[t] = \cos^2[\theta_B/2]G_+e^{-i\omega_B t} + \sin^2[\theta_B/2]G_-e^{i\omega_B t} \\ \beta[t] = e^{i\phi_B}(G_+e^{-i\omega_B t} - G_-e^{i\omega_B t})\sin[\theta_B/2]\cos[\theta_B/2] . \end{cases} \tag{2.62}$$

By using Eq. (2.62), the wave function under the arbitrary magnetic field (2.52) becomes

$$\begin{aligned} \psi_B[t] = &G_+\cos[\theta_B/2]\left(\cos[\theta_B/2]|\Uparrow\rangle + e^{i\phi_B}\sin[\theta_B/2]|\Downarrow\rangle\right)e^{-i\omega_B t} \\ &+ G_-\sin[\theta_B/2]\left(\sin[\theta_B/2]|\Uparrow\rangle - e^{i\phi_B}\cos[\theta_B/2]|\Downarrow\rangle\right)e^{i\omega_B t} . \end{aligned} \tag{2.63}$$

By putting Eq. (2.61) into Eq. (2.62), we obtain the time development of the wave function,

$$\begin{pmatrix} \alpha[t] \\ \beta[t] \end{pmatrix} = \left(\cos \omega_B t\, I - i \sin \omega_B t \begin{pmatrix} \cos \theta_B & e^{i\phi_B} \sin \theta_B \\ e^{-i\phi_B} \sin \theta_B & -\cos \theta_B \end{pmatrix} \right) \begin{pmatrix} \alpha[0] \\ \beta[0] \end{pmatrix}. \quad (2.64)$$

From this, the general wave function (2.63) can be written as follows:

$$\psi_B[t] = \alpha[t]|\Uparrow\rangle + \beta[t]|\Downarrow\rangle = \cdots$$

$$= \left(\frac{\left(\cos \tfrac{\theta}{2}\alpha[0] + e^{-i\phi} \sin \tfrac{\theta}{2}\beta[0] \right) \left(\cos \tfrac{\theta}{2}|\Uparrow\rangle + e^{i\phi} \sin \tfrac{\theta}{2}|\Downarrow\rangle \right) e^{-i\omega t}}{+ \left(e^{i\phi} \sin \tfrac{\theta}{2}\alpha[0] - \cos \tfrac{\theta}{2}\beta[0] \right) \left(e^{-i\phi} \sin \tfrac{\theta}{2}|\Uparrow\rangle - \cos \tfrac{\theta}{2}|\Downarrow\rangle \right) e^{i\omega t}} \right).$$

$$(2.65)$$

2.7.1 Energy Eigenstate

In order to obtain the positive energy eigenstate, we require the coefficient of the $e^{i\omega t}$ term in Eq. (2.63) to be 0,

$$G_- = \alpha[0] - e^{-i\phi_B} \cot[\theta_B/2]\beta[0] = 0 . \quad (2.66)$$

Therefore,

$$\beta[0] = e^{i\phi_B} \tan[\theta_B/2]\alpha[0] . \quad (2.67)$$

The normalization condition is

$$|\alpha[0]|^2 + |\beta[0]|^2 = \frac{1}{\cos^2[\theta_B/2]}|\alpha[0]|^2 = 1 \quad (2.68)$$

that leads

$$\alpha[0] = e^{i\delta_+} \cos[\theta_B/2], \quad \beta[0] = e^{i(\delta_+ + \phi_B)} \sin[\theta_B/2], \quad (2.69)$$

where δ_+ is an arbitrary real constant. In this case

$$G_+ = e^{i\delta_+}(\cos[\theta_B/2] + \tan[\theta_B/2] \sin[\theta_B/2]) = \frac{e^{i\delta_+}}{\cos[\theta_B/2]} \quad (2.70)$$

and the energy eigenstate with energy $E = +\omega_B$ is

$$|\psi_+\rangle = e^{i\delta_+}(\cos[\theta_B/2]|\Uparrow\rangle + e^{i\phi_B} \sin[\theta_B/2]|\Downarrow\rangle) . \quad (2.71)$$

Similarly, the energy eigenstate with energy $E = -\omega_B$ is

$$|\psi_-\rangle = e^{i\delta_-}(\sin[\theta_B/2]|\Uparrow\rangle - e^{i\phi_B} \cos[\theta_B/2]|\Downarrow\rangle) . \quad (2.72)$$

Since the spin direction is defined as the direction of the magnetic field which generates an energy shift of $E = \omega_B$, the spin wave function which points to the direction (θ, ϕ) is

$$\hat{s}[\theta, \phi] = e^{i\delta} \begin{pmatrix} \cos[\theta/2] \\ e^{i\phi} \sin[\theta/2] \end{pmatrix} \tag{2.73}$$

δ is often chosen as $\delta = -\phi/2$ to make the spin wave function in a symmetric form;

$$\hat{s}[\theta, \phi] = \begin{pmatrix} e^{-i(\phi/2)} \cos[\theta/2] \\ e^{i(\phi/2)} \sin[\theta/2] \end{pmatrix}. \tag{2.74}$$

2.7.2 Oscillation and Precession

If the spin direction at time $t = 0$ is $\psi_B[0] = |\Uparrow\rangle$, the initial condition is

$$\psi_B[0] = \alpha[0]|\Uparrow\rangle + \beta[0]|\Downarrow\rangle = |\Uparrow\rangle \tag{2.75}$$

and the initial amplitudes are determined as $(\alpha[0], \beta[0]) = (1, 0)$. At an arbitrary time t, the amplitude of the $|\Uparrow\rangle$ and $|\Downarrow\rangle$ can be obtained by Eq. (2.64) and the wave function in this condition is

$$\psi_B[t] = (\cos \omega_B t - i \cos \theta_B \sin \omega_B t)|\Uparrow\rangle - i \sin \theta_B e^{-i\phi_B} \sin \omega_B t|\Downarrow\rangle. \tag{2.76}$$

The probability that the spin is $|\Downarrow\rangle$ state at time t is

$$P_{\Uparrow \rightarrow \Downarrow}[t] = |\langle\Downarrow | \psi[t]\rangle|^2 = \sin^2 \theta_B \sin^2 \omega_B t. \tag{2.77}$$

Additional factor $\sin^2 \theta_B$ came in the oscillation probability. The oscillation probabilities (2.49) and (2.36) correspond to the case $\theta_B = \pi/2$ in Eq. (2.77).

In order to understand the physical meaning of the new factor, we investigate the state (2.76). The time dependent spin direction vector corresponding to the state (2.76) is, from relation (A.4),

$$\begin{aligned} \vec{e}_s[t] &= \hat{\psi}_B^\dagger[t]\vec{\sigma}\psi_B[t] \\ &= 2 \sin \theta_B \sin \omega_B t(\cos \phi_B \cos \theta_B \sin \omega_B t + \sin \phi_B \cos \omega_B t)\vec{e}_x \\ &\quad + 2 \sin \theta_B \sin \omega_B t(\sin \phi_B \cos \theta_B \sin \omega_B t - \cos \phi_B \cos \omega_B t)\vec{e}_y \\ &\quad + (1 - 2 \sin^2 \theta_B \sin^2 \omega_B t)\vec{e}_z. \end{aligned} \tag{2.78}$$

This indicates that the spin is precessing as shown in Fig. 2.10. The spin direction rotates around the axis (θ_B, ϕ_B), staring from $\vec{S} = (0, 0, 1)$. The amplitude of the oscillation along the z-axis changes as much as $\sin^2 \theta_B$.

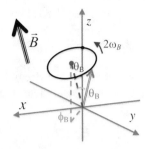

Fig. 2.10 Spin precession under the magnetic field $\vec{B} = B(\sin\theta_B \cos\phi_B, \sin\theta_B \sin\phi_B, \cos\theta_B)$ and the initial condition $\psi_B[0] = |\Uparrow\rangle$. The spin direction rotates around the axis (θ_B, ϕ_B), staring from $\vec{S} = (0, 0, 1)$. The amplitude of the oscillation along the z-axis changes as much as $\sin^2\theta_B$

2.8   If Electron Mass is Included

In quantum mechanics, the equation of motion of two bases states is generally written as

$$i\frac{d\psi[t]}{dt} = H\psi[t]. \tag{2.79}$$

Using the time derivative term, the wave function at infinitesimal time after, $t + \delta t$, can be written as

$$\psi[t + \delta t] \sim \psi[t] + \dot{\psi}[t]\delta t = \psi[t] - iH\psi[t]\delta t. \tag{2.80}$$

The probability of the existence at $t + \delta t$ is

$$P[t + \delta t] = |\psi[t + \delta t]|^2 \sim |\psi[t]|^2 + i\psi^\dagger[t](H^\dagger - H)\psi[t]. \tag{2.81}$$

If we require that the probability conserves, $P[t + \delta t] = P[t]$, H has to be Hermitian,

$$H^\dagger = H. \tag{2.82}$$

Therefore, the most general form of the 2×2 Hamiltonian is

$$H = \begin{pmatrix} a & c^* \\ c & b, \end{pmatrix} \tag{2.83}$$

where a and b are real and c is complex parameters. There are four free parameters. While the Pauli equation includes only three free parameters. Therefore, the Pauli equation has less degree of freedom and another parameter can be included to express the most general two-body phenomena.

If the electron mass m_e is taken into account as a common energy shift, the Hamiltonian of the spin under a magnetic field becomes

$$H = H_0 + H_B = \begin{pmatrix} m_e & 0 \\ 0 & m_e \end{pmatrix} + \mu_B \begin{pmatrix} B_z & B_- \\ B_+ & -B_z \end{pmatrix} = \begin{pmatrix} m_e + \mu_B B_z & \mu_B B_- \\ \mu_B B_+ & m_e - \mu_B B_z. \end{pmatrix} \tag{2.84}$$

We can relate the general parameters in Eq. (2.83) to the parameters in Eq. (2.84) as

$$a = m_e + \mu_B B_z, \quad b = m_e - \mu_B B_z, \quad c = \mu_B B_+. \tag{2.85}$$

This means any two-body phenomena without decay has a correspondence to the spin motion of spin-1/2 electron in a magnetic field. Using Eq. (2.84), the equation of motion can be written as

$$i\dot{\psi} = (H_0 + H_B)\psi. \tag{2.86}$$

If we write $\psi = \psi' e^{-imt}$, the equation of motion (2.86) becomes

$$i\dot{\psi}' e^{-imt} + m\psi' e^{-imt} = (H_0 + H_B)\psi' e^{-imt} = m\psi' e^{-imt} + H_B\psi' e^{-imt}. \tag{2.87}$$

Eliminating the common terms in the first and third equations,

$$i\dot{\psi}' = H_B\psi'. \tag{2.88}$$

This equation is the same as Eq. (2.53). Therefore, ψ' is equivalent to Eq. (2.65).

Physically the mass shifts the total energy ($E = \pm\omega \rightarrow E = m \pm \omega$) and all aspects of the spin-magnetic interactions are not affected by it. Most of the subjects dealt in this textbook can start from this wave function formula.

Any 2×2 Hamiltonian (2.83) can be separated to the terms correspond to H_0 and H_B,

$$H = \begin{pmatrix} a & c^* \\ c & b \end{pmatrix} = \begin{pmatrix} (a+b)/2 & 0 \\ 0 & (a+b)/2 \end{pmatrix} + \begin{pmatrix} (a-b)/2 & c^* \\ c & -(a-b)/2 \end{pmatrix} \tag{2.89}$$

The relation between $a \sim c$ and the parameters used in Eq. (2.53) can be obtained using the relation (2.85), as,

$$m = \frac{a+b}{2}, \quad \tan\theta = \frac{|c|}{a-b}, \quad e^{i\phi} = \frac{c}{|c|}, \quad \omega = \sqrt{\left(\frac{a-b}{2}\right)^2 + |c|^2}, \tag{2.90}$$

Fig. 2.11 Mixing triangle shows the relation between (a, b, c) and (θ, ω). **a** For the case $a > b$ ($0 \leq \theta < \pi/2$). **b** For the case $a < b$ ($\pi/2 < \theta < \pi$)

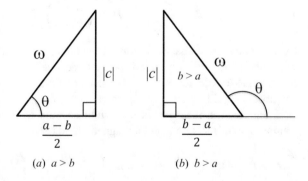

where the domains of θ and ϕ are $0 \leq \theta < \pi$ and $0 \leq \phi < 2\pi$. The triangle, called *mixing triangle*, shown in Fig. 2.11 is useful to memorize the relations between (a, b, c) and (θ, ω). In case $(a > b)$, the domain of θ is $(0 \leq \theta < \pi/2)$ as shown in Fig. 2.11a, while in case $(a < b)$, the domain of θ is $(\pi/2 < \theta < \pi)$ as shown in Fig. 2.11b. In two-body oscillation phenomena without decay, ϕ does not contribute to physical effects.

Reference

1. Halzen, F., Martin, A.D.: Quarks and Leptons. Wiley, Hoboken (1984)

Hydrogen Hyperfine Splitting and HI Line

3

3.1 Introduction

The hydrogen atom is made of a proton and an electron. They are spin-1/2 fermions and have magnetic moments (MM). The magnetic moment-magnetic moment (MM-MM) interaction takes place in the hydrogen atom. The energy scale of this interaction is $\mu_p\mu_e/a_B^3 \sim O[1]\mu\,eV$, where μ_p and μ_e are magnetic moments of proton and electron, and a_B is the Bohr radius. This is very much smaller than the electrostatic potential energy, $\alpha/a_B \sim O[10]\,eV$. However, the MM-MM interaction causes the hydrogen HI line and plays a very important role in astronomy. In this chapter, MM-MM interactions between two spins are discussed and we will understand the mechanism of the hydrogen HI line quantum mechanically.

3.2 Spin Structure of the Hydrogen Atom

Figure 3.1 shows the structure of the magnetic moment of the hydrogen atom. The intrinsic magnetic moments of the electron and the proton are

$$\mu_e = -\mu_B = -\frac{e}{2m_e}, \qquad \mu_p = \kappa_p \frac{e}{2m_p} \sim 2.793\frac{e}{2m_p} \equiv 2.793\mu_N , \qquad (3.1)$$

where μ_B is the Bohr magneton ($\mu_B = e/2m_e = 5.788 \times 10^{-5}\,eV/T$) and μ_N is the nuclear magneton ($\mu_N = e/2m_p = 3.152 \times 10^{-8}\,eV/T$). The relation among the magnetic moment (μ), charge (q), and mass (m) of a spin-1/2 elementary particle,

$$\mu = \frac{q}{2m} , \qquad (3.2)$$

is a universal relation which comes from the relativistic quantum mechanics. The proton magnetic moment deviates from the relation because the proton has internal

© Springer Nature Switzerland AG 2021
F. Suekane, *Quantum Oscillations*, Lecture Notes in Physics 985,
https://doi.org/10.1007/978-3-030-70527-5_3

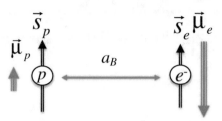

Fig. 3.1 Hydrogen spin structure and magnetic moments. $\vec{S}$ shows spin vector, $\vec{\mu}$ shows magnetic moment and a_B is the Bohr radius ($\sim$0.53Å). Because the electron charge is negative, the direction of its magnetic moment $\vec{\mu}_e$ is opposite to the direction of its spin $\vec{S}_e$. $|\vec{\mu}_e| \sim 5.8 \times 10^{-5}$ eV/T, $|\vec{\mu}_p| \sim 8.8 \times 10^{-8}$ eV/T and $\Delta E \sim (|\vec{\mu}_e||\vec{\mu}_p|)/a_B^3 \sim O[1]\,\mu$eV

structure and is not an elementary particle. This deviation is called proton anomalous magnetic moment and the factor of the deviation is expressed by $\kappa_p (\sim 2.79)$. Since the proton mass is much larger than the electron mass, the magnetic moment of the proton is much smaller than that of the electron,

$$\left|\frac{\mu_p}{\mu_e}\right| = \kappa_p \frac{m_e}{m_p} \sim 1.5 \times 10^{-3} \,. \tag{3.3}$$

In classical electrodynamics, the proton magnetic moment produces a magnetic field at a position $\vec{r}$ as[1]

$$\vec{B}_p = \frac{1}{4\pi r^3}\left(3(\vec{\mu}_p \cdot \hat{r})\hat{r} - \vec{\mu}_p\right) + \frac{2}{3}\vec{\mu}_p\delta^3[\vec{r}] \,, \tag{3.4}$$

where $\vec{r}$ is the relative position from the proton and $\hat{r}$ is the unit vector of the direction of $\vec{r}$. The first term is the magnetic field outside the proton and the second term corresponds to the one *inside* the proton.[2] The electron in the magnetic field $\vec{B}_p$ obtains the potential energy,

$$V_M[\vec{r}] = -\vec{\mu}_e \cdot \vec{B}_p \sim -\frac{1}{4\pi r^3}\left(3(\vec{\mu}_p \cdot \hat{r})(\vec{\mu}_e \cdot \hat{r}) - (\vec{\mu}_p \cdot \vec{\mu}_e)\right) - \frac{2}{3}(\vec{\mu}_p \cdot \vec{\mu}_e)\delta^3[\vec{r}] \,. \tag{3.5}$$

The average energy shift of the electron due to the MM-MM interaction is

$$E_M = \int V_M[\vec{r}]\rho_e[\vec{r}]d^3\vec{r} \,, \tag{3.6}$$

where $\rho_e[\vec{r}]$ is probability density of the electron. Now we consider the electron is orbiting in the $1S$ state. In this case, the wave function and the probability density

---

[1] For example, Eq. (5.58) of [1].
[2] For intuitive description, see [2].

of the electron are

$$\psi_{1S}[\vec{r}] = \frac{1}{\sqrt{\pi a_B^3}} e^{-r/a_B} \quad \rightarrow \quad \rho_e[\vec{r}] = |\psi_{1S}[\vec{r}]|^2 = \frac{1}{\pi a_B^3} e^{-2r/a_B} , \tag{3.7}$$

where Bohr radius is $a_B = 1/\alpha m_e \sim 0.53\text{Å}$. The integration of the first term of Eq. (3.5) vanishes and the energy shift (3.6) becomes

$$E_M = -\frac{2}{3\pi a_B^3} (\vec{\mu}_e \cdot \vec{\mu}_p) . \tag{3.8}$$

The size of the energy shift is

$$-\frac{2\mu_e\mu_p}{3\pi a_B^3} = \frac{2\kappa_p \alpha^4 m_e^2}{3m_p} \sim 1.47\,\mu\text{eV} . \tag{3.9}$$

This is only 10^{-7} of the potential energy of the $1S$ state ($E_{1S} = -13.6\,\text{eV}$).

In order to quantize the relation (3.8), we replace

$$\vec{\mu}_e \rightarrow \mu_e \vec{\sigma}_e, \quad \vec{\mu}_p \rightarrow \mu_p \vec{\sigma}_p, \quad E_M \rightarrow H_M, \tag{3.10}$$

and make an equation of motion applying a wave function ψ_M to H_M,

$$i\frac{d}{dt}\psi_M = H_M\psi_M = -\frac{2\mu_e\mu_p}{3\pi a_B^3}(\vec{\sigma}_e \cdot \vec{\sigma}_p)\psi_M \equiv A_H(\vec{\sigma}_e \cdot \vec{\sigma}_p)\psi_M . \tag{3.11}$$

The above derivation is empirical but it is also possible to derive the same MM-MM interaction term from low energy approximation of QED for electromagnetic interactions of two fermions.[3]

In Eq. (3.11), the wave function ψ_M is a combination of the spin wave functions of the electron and proton,

$$|s_e\rangle = \alpha_e|\Uparrow\rangle_e + \beta_e|\Downarrow\rangle_e, \quad |s_p\rangle = \alpha_p|\Uparrow\rangle_p + \beta_p|\Downarrow\rangle_p . \tag{3.12}$$

Then ψ_M can be written as

$$\begin{aligned}|\psi_M[t]\rangle = |s_e\rangle|s_p\rangle &= (\alpha_e[t]|\Uparrow\rangle_e + \beta_e[t]|\Downarrow\rangle_e)(\alpha_p[t]|\Uparrow\rangle_p + \beta_p[t]|\Downarrow\rangle_p) \\ &\equiv C_{\Uparrow\Uparrow}[t]|\Uparrow\Uparrow\rangle + C_{\Uparrow\Downarrow}[t]|\Uparrow\Downarrow\rangle + C_{\Downarrow\Uparrow}[t]|\Downarrow\Uparrow\rangle + C_{\Downarrow\Downarrow}[t]|\Downarrow\Downarrow\rangle ,\end{aligned} \tag{3.13}$$

where $|\Uparrow\Downarrow\rangle \equiv |\Uparrow\rangle_e|\Downarrow\rangle_p$ and $C_{\Uparrow\Downarrow} \equiv \alpha_e\beta_p$ and so on.

---

[3]For example, Sect. 6.5, Eq. (11) of [3].

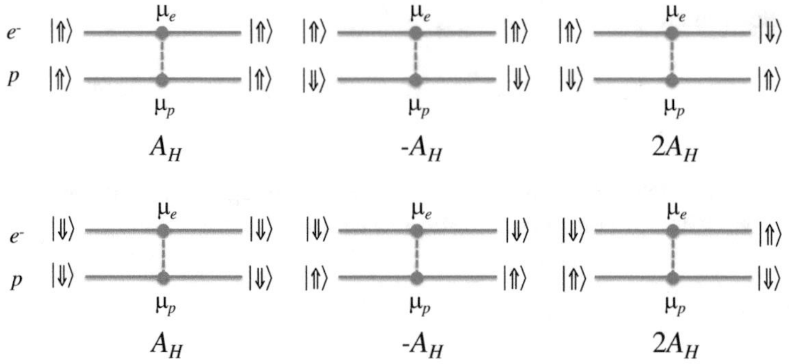

Fig. 3.2 Spin transition amplitudes due to the MM-MM interaction

The last term of Eq. (3.11) can be written as

$$
\begin{aligned}
(\vec{\sigma}_e \cdot \vec{\sigma}_p)\psi_M &\rightarrow (\vec{\sigma}_e|s_e)) \cdot (\vec{\sigma}_p|s_p)) \\
&= \sigma_e^x|s_e\rangle\sigma_p^x|s_p\rangle + \sigma_e^y|s_e\rangle\sigma_p^y|s_p\rangle + \sigma_e^z|s_e\rangle\sigma_p^z|s_p\rangle .
\end{aligned}
\tag{3.14}
$$

The Pauli matrices give the relations,

$$
\begin{aligned}
\sigma_x|\Uparrow\rangle = |\Downarrow\rangle, \quad \sigma_y|\Uparrow\rangle = i|\Downarrow\rangle, \quad \sigma_z|\Uparrow\rangle = |\Uparrow\rangle \\
\sigma_x|\Downarrow\rangle = |\Uparrow\rangle, \quad \sigma_y|\Downarrow\rangle = -i|\Uparrow\rangle, \quad \sigma_z|\Downarrow\rangle = -|\Downarrow\rangle .
\end{aligned}
\tag{3.15}
$$

Therefore,

$$
\begin{aligned}
\vec{\sigma}_e \cdot \vec{\sigma}_p|\Uparrow\Uparrow\rangle = |\Uparrow\Uparrow\rangle, \quad \vec{\sigma}_e \cdot \vec{\sigma}_p|\Uparrow\Downarrow\rangle = 2|\Downarrow\Uparrow\rangle - |\Uparrow\Downarrow\rangle, \\
\vec{\sigma}_e \cdot \vec{\sigma}_p|\Downarrow\Uparrow\rangle = 2|\Uparrow\Downarrow\rangle - |\Downarrow\Uparrow\rangle, \quad \vec{\sigma}_e \cdot \vec{\sigma}_p|\Downarrow\Downarrow\rangle = |\Downarrow\Downarrow\rangle .
\end{aligned}
\tag{3.16}
$$

Using these relations, Eq. (3.11) becomes

$$
\begin{aligned}
&\dot{C}_{\Uparrow\Uparrow}|\Uparrow\Uparrow\rangle + \dot{C}_{\Uparrow\Downarrow}|\Uparrow\Downarrow\rangle + \dot{C}_{\Downarrow\Uparrow}|\Downarrow\Uparrow\rangle + \dot{C}_{\Downarrow\Downarrow}|\Downarrow\Downarrow\rangle \\
&= -iA_H(C_{\Uparrow\Uparrow}|\Uparrow\Uparrow\rangle + (2C_{\Downarrow\Uparrow} - C_{\Uparrow\Downarrow})|\Uparrow\Downarrow\rangle + (2C_{\Uparrow\Downarrow} - C_{\Downarrow\Uparrow})|\Downarrow\Uparrow\rangle + C_{\Downarrow\Downarrow}|\Downarrow\Downarrow\rangle) .
\end{aligned}
\tag{3.17}
$$

Comparing the coefficients of the four basis vectors, we obtain the following set of differential equations for the MM-MM interactions:

$$
\frac{d}{dt}\begin{pmatrix} C_{\Uparrow\Uparrow} \\ C_{\Uparrow\Downarrow} \\ C_{\Downarrow\Uparrow} \\ C_{\Downarrow\Downarrow} \end{pmatrix} = -iA_H \begin{pmatrix} 1 & 0 & 0 & 0 \\ 0 & -1 & 2 & 0 \\ 0 & 2 & -1 & 0 \\ 0 & 0 & 0 & 1 \end{pmatrix} \begin{pmatrix} C_{\Uparrow\Uparrow} \\ C_{\Uparrow\Downarrow} \\ C_{\Downarrow\Uparrow} \\ C_{\Downarrow\Downarrow} \end{pmatrix} .
\tag{3.18}
$$

Figure 3.2 shows the spin transition amplitudes that correspond to the differential equations (3.18).

The first and the fourth equations in Eq. (3.18) can be directly solved as

$$C_{\uparrow\uparrow}[t] = ae^{-iA_H t}, \quad C_{\Downarrow\Downarrow}[t] = be^{-iA_H t}. \tag{3.19}$$

The second and third equations can be solved using the add-subtract method as

$$C_{\uparrow\Downarrow}[t] = pe^{-iA_H t} + qe^{3iA_H t}, \quad C_{\Downarrow\uparrow}[t] = pe^{-iA_H t} - qe^{3iA_H t}. \tag{3.20}$$

The general wave function is then

$$\begin{aligned}
|\psi_M[t]\rangle &= ae^{-iA_H t}|\uparrow\uparrow\rangle + be^{-iA_H t}|\Downarrow\Downarrow\rangle + pe^{-iA_H t}(|\uparrow\Downarrow\rangle + |\Downarrow\uparrow\rangle) \\
&\quad + qe^{3iA_H t}(|\uparrow\Downarrow\rangle - |\Downarrow\uparrow\rangle).
\end{aligned} \tag{3.21}$$

The energy eigenstates are

$$\begin{aligned}
|\psi_I[t]\rangle &= e^{-iA_H t}|\uparrow\uparrow\rangle, \quad |\psi_{II}[t]\rangle = e^{-iA_H t}|\Downarrow\Downarrow\rangle \\
|\psi_{III}[t]\rangle &= e^{-iA_H t}\frac{|\uparrow\Downarrow\rangle + |\Downarrow\uparrow\rangle}{\sqrt{2}} \equiv e^{-iA_H t}|+\rangle \\
|\psi_{IV}[t]\rangle &= e^{3iA_H t}\frac{|\uparrow\Downarrow\rangle - |\Downarrow\uparrow\rangle}{\sqrt{2}} \equiv e^{i3A_H t}|-\rangle.
\end{aligned} \tag{3.22}$$

The three states have the same energy A_H and one state has energy $-3A_H$. The average energy due to the MM-MM interaction is 0. Figure 3.3 shows the level structure generated by the MM-MM interaction. The expected energy split is

$$\Delta E_D^H = 4A_H = -\frac{8\mu_e\mu_p}{3\pi a_B^3} = 2.7928473446\frac{8}{3}\alpha^4\frac{m_e^2}{m_p} \sim 5.8774\,\mu\text{eV}. \tag{3.23}$$

On the other hand, the measured value is

$$\Delta E_{hfs} = 5.87433\,\mu\text{eV}. \tag{3.24}$$

The general wave function can be expressed by the energy eigenstate as

$$|\psi_M[t]\rangle = C_I|\psi_I[t]\rangle + C_{II}|\psi_{II}[t]\rangle + C_{III}|\psi_{III}[t]\rangle + C_{IV}|\psi_{IV}[t]\rangle. \tag{3.25}$$

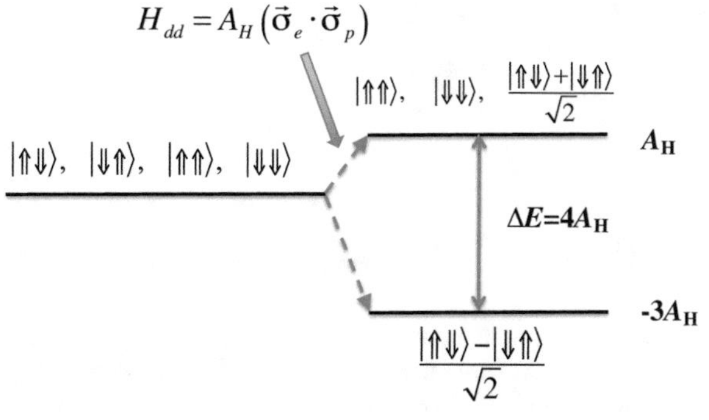

Fig. 3.3 Hyperfine level splitting

3.2.1 Oscillation Between $|\Uparrow\Downarrow\rangle$ and $|\Downarrow\Uparrow\rangle$ States

In principle, it is possible to polarize a proton and attach a polarized electron on it to form a hydrogen with the spin structure; $|\Uparrow\Downarrow\rangle$. Let's see what happens after such hydrogen is formed. Physically the above manipulation corresponds to fixing the initial condition. That is, at $t = 0$, the system is pure $|\Uparrow\Downarrow\rangle$ state; $\psi_M[0] = |\Uparrow\Downarrow\rangle$. From Eq. (3.25)

$$\psi_M[0] = C_I\psi_I[0] + C_{II}\psi_{II}[0] + C_{III}\psi_{III}[0] + C_{IV}\psi_{IV}[0]$$
$$= C_I|\Uparrow\Uparrow\rangle + C_{II}|\Downarrow\Downarrow\rangle + \frac{C_{III} + C_{IV}}{\sqrt{2}}|\Uparrow\Downarrow\rangle + \frac{C_{III} - C_{IV}}{\sqrt{2}}|\Downarrow\Uparrow\rangle = |\Uparrow\Downarrow\rangle \,.$$
(3.26)

Therefore, the coefficients are determined as

$$C_I = C_{II} = 0, \quad C_{III} = C_{IV} = \frac{1}{\sqrt{2}} \,. \tag{3.27}$$

In this case the wave function at time t is

$$\psi_M[t] = \frac{1}{\sqrt{2}}(e^{-iA_H t}|+\rangle + e^{i3A_H t}|-\rangle)$$
$$= e^{iA_H t}(\cos[2A_H t]|\Uparrow\Downarrow\rangle - i\sin[2A_H t]|\Downarrow\Uparrow\rangle) \,. \tag{3.28}$$

This is the wave function for the state which started from $|\Uparrow\Downarrow\rangle$. After a finite time t, $|\Downarrow\Uparrow\rangle$ state is generated with probability,

$$P_{|\Uparrow\Downarrow\rangle\to|\Downarrow\Uparrow\rangle}[t] = |\langle\Downarrow\Uparrow|\psi_M[t]\rangle|^2 = \sin^2[2A_H t] \tag{3.29}$$

and the probability that $|\Uparrow\Downarrow\rangle$ state remains at the time t is

$$P_{|\Uparrow\Downarrow\rangle\to|\Uparrow\Downarrow\rangle}[t] = |\langle\Uparrow\Downarrow|\psi_M[t]\rangle|^2 = \cos^2[2A_H t] \,. \tag{3.30}$$

These probabilities are shown in Fig. 3.4. The probability oscillates with a period, $T = \pi/2A_H = 7 \times 10^{-10}\,\mathrm{s} = 21\,\mathrm{cm}$.

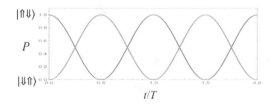

Fig. 3.4 Time dependence of the oscillation probability between $|\Uparrow_e\Downarrow_p\rangle$ and $|\Downarrow_e\Uparrow_p\rangle$ spin states in the hydrogen atom due to the MM-MM interaction. Horizontal axis is time in unit of $T = \pi/2A_H$. Vertical axis is the probability of being in either state

3.3 Hydrogen Magnetic Moment Under External Magnetic Field

If we apply an external magnetic field $\vec{B}$ to the hydrogen atom, the potential energy becomes

$$E_{\mu B} = -(\vec{\mu}_e + \vec{\mu}_p) \cdot \vec{B} . \tag{3.31}$$

We specify the direction of the magnetic field to be along the z-axis; $\vec{B} = (0, 0, B)$ for simplicity. Then we quantize the relation (3.31) and obtain the following equation of motion:

$$i \frac{d}{dt} \psi = -B(\mu_e \sigma_e^z + \mu_p \sigma_p^z)\psi . \tag{3.32}$$

We treat the wave function ψ by the two spin basis as Eq. (3.13). Since

$$\sigma^z | \Uparrow \rangle = | \Uparrow \rangle, \quad \sigma^z | \Downarrow \rangle = -| \Downarrow \rangle , \tag{3.33}$$

the calculation of one of the elements of the right-hand side of Eq. (3.32) is performed as follows:

$$
\begin{aligned}
(\mu_e \sigma_e^z + \mu_p \sigma_p^z)| \Uparrow_e \Downarrow_p \rangle &= \mu_e (\sigma_e^z | \Uparrow \rangle_e)| \Downarrow \rangle_p + \mu_p | \Uparrow \rangle_e (\sigma_p^z | \Downarrow \rangle_p) \\
&= (\mu_e - \mu_p)| \Uparrow_e \Downarrow_p \rangle .
\end{aligned} \tag{3.34}
$$

There are four cases in total,

$$
\begin{cases}
(\mu_e \sigma_e^z + \mu_p \sigma_p^z)| \Uparrow_e \Uparrow_p \rangle = \mu_+ | \Uparrow_e \Uparrow_p \rangle \\
(\mu_e \sigma_e^z + \mu_p \sigma_p^z)| \Uparrow_e \Downarrow_p \rangle = \mu_- | \Uparrow_e \Downarrow_p \rangle \\
(\mu_e \sigma_e^z + \mu_p \sigma_p^z)| \Downarrow_e \Uparrow_p \rangle = -\mu_- | \Downarrow_e \Uparrow_p \rangle \\
(\mu_e \sigma_e^z + \mu_p \sigma_p^z)| \Downarrow_e \Downarrow_p \rangle = -\mu_+ | \Downarrow_e \Downarrow_p \rangle ,
\end{cases} \tag{3.35}
$$

where $\mu_\pm = \mu_e \pm \mu_p$. These equations mean that the outer magnetic field generates the self-transition amplitudes only as shown in Fig. 3.5.

Since the MM-MM interaction intrinsically exists within the hydrogen atom, we have to combine the effects caused by the external magnetic field (Fig. 3.5) and the transition amplitudes of the MM-MM interactions (Fig. 3.2). How is it possible to combine the effects of different processes? The answer is, just add the amplitudes of the same initial states and the same final states. Figure 3.6 shows the transition

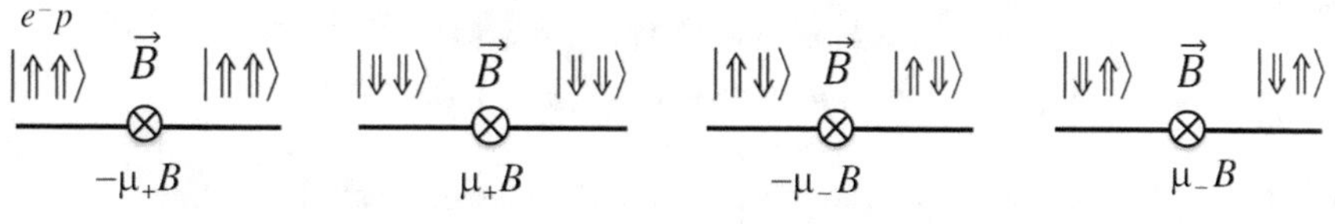

Fig. 3.5 Transition amplitudes for an external magnetic field $\vec{B} = (0, 0, B)$. Note that there is no cross-transition

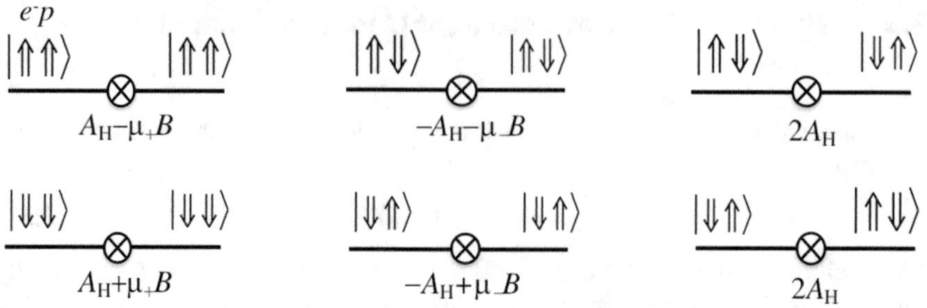

$$e^- p$$

$$
|\!\uparrow\uparrow\rangle \;\;\underset{A_\mathrm{H}-\mu_+B}{\otimes}\;\; |\!\uparrow\uparrow\rangle
\qquad\qquad
|\!\uparrow\Downarrow\rangle \;\;\underset{-A_\mathrm{H}-\mu_-B}{\otimes}\;\; |\!\uparrow\Downarrow\rangle
\qquad\qquad
|\!\uparrow\Downarrow\rangle \;\;\underset{2A_\mathrm{H}}{\otimes}\;\; |\!\Downarrow\uparrow\rangle
$$

$$
|\!\Downarrow\Downarrow\rangle \;\;\underset{A_\mathrm{H}+\mu_+B}{\otimes}\;\; |\!\Downarrow\Downarrow\rangle
\qquad\qquad
|\!\Downarrow\uparrow\rangle \;\;\underset{-A_\mathrm{H}+\mu_-B}{\otimes}\;\; |\!\Downarrow\uparrow\rangle
\qquad\qquad
|\!\Downarrow\uparrow\rangle \;\;\underset{2A_\mathrm{H}}{\otimes}\;\; |\!\uparrow\Downarrow\rangle
$$

Fig. 3.6 Transition amplitude for both MM-MM interaction and external magnetic field $\vec{B}$

amplitudes for both MM-MM interaction and external magnetic field, obtained by adding corresponding amplitudes. As a result, the transition equation becomes

$$
\frac{d}{dt}\begin{pmatrix} C'_{\uparrow\uparrow} \\ C'_{\uparrow\Downarrow} \\ C'_{\Downarrow\uparrow} \\ C'_{\Downarrow\Downarrow} \end{pmatrix}
= -i
\begin{pmatrix}
A_\mathrm{H}-\mu_+B & 0 & 0 & 0 \\
0 & -A_\mathrm{H}-\mu_-B & 2A_\mathrm{H} & 0 \\
0 & 2A_\mathrm{H} & -A_\mathrm{H}+\mu_-B & 0 \\
0 & 0 & 0 & A_\mathrm{H}+\mu_+B
\end{pmatrix}
\begin{pmatrix} C'_{\uparrow\uparrow} \\ C'_{\uparrow\Downarrow} \\ C'_{\Downarrow\uparrow} \\ C'_{\Downarrow\Downarrow} \end{pmatrix},
\tag{3.36}
$$

where C'_X are the coefficients of the basic states for the case both MM-MM and external magnetic field are working. The first and fourth equations in Eq. (3.36) can directly be solved as

$$
C'_{\uparrow\uparrow}[t] = a e^{-i(A_\mathrm{H}-\mu_+B)t}, \qquad C'_{\Downarrow\Downarrow}[t] = b e^{-i(A_\mathrm{H}+\mu_+B)t}.
\tag{3.37}
$$

The corresponding energy eigenstates are

$$
|\psi'_\mathrm{I}[t]\rangle = e^{-i(A_\mathrm{H}-\mu_+B)t}|\!\uparrow\uparrow\rangle, \qquad |\psi'_\mathrm{II}[t]\rangle = e^{-i(A_\mathrm{H}+\mu_+B)t}|\!\Downarrow\Downarrow\rangle.
\tag{3.38}
$$

This means that the structure of these energy eigenstates are the same as the ones without the external magnetic field,

$$
|\psi'_\mathrm{I}\rangle = |\psi_\mathrm{I}\rangle = |\!\uparrow\uparrow\rangle, \qquad |\psi'_\mathrm{II}\rangle = |\psi_\mathrm{II}\rangle = |\!\Downarrow\Downarrow\rangle.
\tag{3.39}
$$

The second and the third equations of Eq. (3.36) are simultaneous differential equations. The energy eigenstates and energies from the equations are, from Eqs. (B.35),

$$
\begin{cases}
|\psi'_\mathrm{III}\rangle = (\cos\theta_{\mu B}|\!\uparrow\Downarrow\rangle + \sin\theta_{\mu B}|\!\Downarrow\uparrow\rangle) e^{-i(-A_\mathrm{H}+\omega)t} \\
|\psi'_\mathrm{IV}\rangle = (\sin\theta_{\mu B}|\!\uparrow\Downarrow\rangle - \cos\theta_{\mu B}|\!\Downarrow\uparrow\rangle) e^{-i(-A_\mathrm{H}-\omega)t},
\end{cases}
\tag{3.40}
$$

where the mixing angle $\theta_{\mu B}$ and the energy shift ω are expressed as follows:

$$
\tan 2\theta_{\mu B} = \frac{2A_\mathrm{H}}{-\mu_- B}, \qquad \omega = \sqrt{(\mu_- B)^2 + 4A_\mathrm{H}^2}.
\tag{3.41}
$$

Fig. 3.7 Mixing triangle of a hydrogen under the magnetic field B

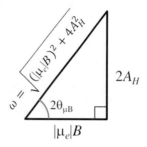

Since $|\mu_e| \sim 700\mu_p$, $\mu_\pm$ can be approximated as $\mu_\pm \sim \mu_e(< 0)$ as mentioned before. Finally the energy eigenstates are expressed as

$$
\begin{cases}
|\psi_I'[t]\rangle = |\uparrow\uparrow\rangle e^{-i(A_H + |\mu_e|B)t} \\
|\psi_{II}'[t]\rangle = |\Downarrow\Downarrow\rangle e^{-i(A_H - |\mu_e|B)t} \\
|\psi_{III}'[t]\rangle = (\cos\theta_{\mu B}|\uparrow\Downarrow\rangle + \sin\theta_{\mu B}|\Downarrow\uparrow\rangle) e^{i(A_H - \omega)t} \\
|\psi_{IV}'[t]\rangle = (\sin\theta_{\mu B}|\uparrow\Downarrow\rangle - \cos\theta_{\mu B}|\Downarrow\uparrow\rangle) e^{i(A_H + \omega)t},
\end{cases}
\tag{3.42}
$$

where

$$
\tan 2\theta_{\mu B} = \frac{2A_H}{|\mu_e|B}, \quad \omega = \sqrt{(|\mu_e|B)^2 + 4A_H^2}.
\tag{3.43}
$$

The corresponding mixing triangle is shown in Fig. 3.7.

The dependence of the energy levels and the structure of the wave function of the mass eigenstate in the external magnetic field B is shown in Fig. 3.8. The mixing angle $\theta_{\mu B}$ depends on the external magnetic field B and accordingly, the spin structures of the energy eigenstates (ψ_{III}', ψ_{IV}') change depending on B. On the other hand, the spin structures of (ψ_I', ψ_{II}') do not change. This is a good example to see the dependence of the mass eigenstate structure on the transition amplitudes.

We consider a case that the potential energy generated by the applied magnetic field is small so as to $|\mu_e|B \ll A_H$ and parametrize the following ratio as ϵ,

$$
\epsilon \equiv \frac{|\mu_e|B}{A_H} \ll 1.
\tag{3.44}
$$

Since ϵ is small, the mixing angle is approximated as

$$
\theta_{\mu B} \sim \frac{\pi}{4} - \frac{\epsilon}{4}.
\tag{3.45}
$$

Accordingly, the energy eigenstates (3.40) are approximated as

$$
\begin{aligned}
|\psi_{III}'\rangle &\sim |\psi_{III}\rangle + \frac{\epsilon}{4}|\psi_{IV}\rangle \\
|\psi_{IV}'\rangle &\sim |\psi_{IV}\rangle - \frac{\epsilon}{4}|\psi_{III}\rangle.
\end{aligned}
\tag{3.46}
$$

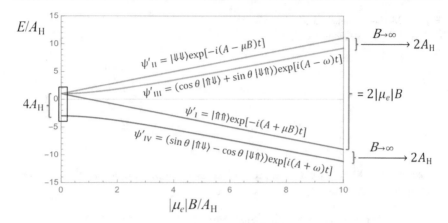

Fig. 3.8 Relation of the energies (E'_{I}, E'_{II}, E'_{III}, E'_{IV}) and the external magnetic field (B). Vertical axis is the potential energy in E/A_{H} units. Horizontal axis is the strength of the external magnetic field in $|\mu_e|B/A_{\mathrm{H}}$ units. The spin structures of the energy eigenstates (ψ'_{III}, ψ'_{IV}) change depending on B but that of (ψ'_{I}, ψ'_{II}) do not change. Properties in the blank box region at $|\mu_e|B \sim 0$ will be discussed later

This means that due to the external magnetic field, the energy eigenstates $|\psi_{\mathrm{III}}\rangle$ and $|\psi_{\mathrm{IV}}\rangle$ mix and become new energy eigenstates. On the other hand, the changes of E'_{III} and E'_{IV} due to the external magnetic field are small,

$$E'_{\mathrm{III}} = E_{\mathrm{III}}(1 + O[\epsilon^2])$$
$$E'_{\mathrm{IV}} = E_{\mathrm{IV}}(1 + O[\epsilon^2]) . \tag{3.47}$$

Figure 3.9 shows the energy separation due to the internal MM-MM interaction and weak external magnetic field. This energy spread pattern is exactly the same as the energies of spin-0 particle with static energy $-3A_{\mathrm{H}}$, and spin-1 particle with static energy $+A_{\mathrm{H}}$ and magnetic moment μ_+, in the external magnetic field. This means that the spin-0 and spin-1 states are constructed by combining two spin-1/2 states. Symbolically it is written as

$$|1/2\rangle \times |1/2\rangle = |0\rangle + |1\rangle. \tag{3.48}$$

The hydrogen atom is a good example to understand an essential difference between quantum mechanics and classical mechanics. From the point of view of classical electrodynamics, the magnitude of the magnetic moment of the ground state hydrogen atom is almost μ_e no matter what structure the hydrogen atom has, because μ_p is much smaller than $|\mu_e|$ and can be ignored. However, observations show that the magnetic moment of the hydrogen's ground state (ψ_{IV}) is 0. How can it be understood?

The spin structure of the spin-0 ground state is $|\psi_{\mathrm{IV}}\rangle = (|\Uparrow_e \Downarrow_p\rangle - |\Downarrow_e \Uparrow_p\rangle)/\sqrt{2}$. In this ground state, the $|\Uparrow_e \Downarrow_p\rangle$ and $|\Downarrow_e \Uparrow_p\rangle$ oscillate with each other due to the MM-MM interaction as shown in Fig. 3.4. On the other hand, if we want to measure

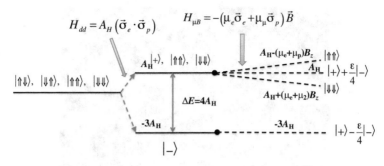

Fig. 3.9 Energy level under weak external magnetic field B. $|\pm\rangle = (|\Uparrow\Downarrow\rangle \pm |\Downarrow\Uparrow\rangle)/\sqrt{2}$ are used

the properties of the $|\psi_{IV}\rangle$ state, we have to separate $|\psi_{IV}\rangle$ from $|\psi_I\rangle$. We separate these two systems making use of the difference of their energy levels; $4A_H$. In order to distinguish the two energy levels, we need energy resolution δE to be better than the energy difference; $\delta E \ll 4A_H$. However, due to the uncertainty principle, if we can determine the energy with precision δE, we cannot have the timing resolution better than $\delta t \sim 1/\delta E \gg 1/4A_H \sim T/(2\pi)$, where T is the oscillation period defined just below Eq. (3.30). This means we cannot tell which state, either $|\Uparrow_e\Downarrow_p\rangle$ or $|\Downarrow_e\Uparrow_p\rangle$, ψ_{IV} is, when we measure the effect of the external magnetic field because they oscillate each other with a frequency quicker than our time resolution.

If we try to measure the magnetic moment of ψ_{IV} state, we apply a magnetic field and measure the caused energy shift. However, the $|\Uparrow_e\Downarrow_p\rangle$ and $|\Downarrow_e\Uparrow_p\rangle$ react with opposite sign amplitude because the relative directions of the magnetic moments are opposite. Since we cannot tell whether the reaction was $|\Uparrow_e\Downarrow_p\rangle$ origin or $|\Downarrow_e\Uparrow_p\rangle$ origin, we have to add the amplitudes and take the absolute square to calculate the probability. Since the amplitudes have the same magnitude but opposite sign, the sum results in 0. That means the ψ_{IV} state does not react to the external magnetic field and it is the definition of spin-0 system.

3.4 Hydrogen 21 cm Line

The ground state hydrogen atom absorbs 5.9 μeV photon, corresponding to 21 cm wavelength. The 21 cm microwave line called HI (pronounced as "H one") line has been used extensively in astronomy. By observing the HI line it is possible to measure the density of the interstellar hydrogen and its speed of sightline direction from its Doppler shift. The HI line was firstly observed in 1951.

Existence of dark matter was motivated from the observation of the HI line. Figure 3.10 shows the rotation speed of hydrogen atom as a function of the distance from the galaxy center measured by the Doppler shift of the HI line. The observed speed pattern deviates from the expectation from the visible mass distribution of the galaxy. This observation indicates that there is an unknown mass called "dark matter" in the galaxy.

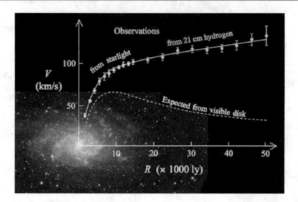

Fig. 3.10 Rotation speed of a galaxy as a function of the distance from the galaxy center measured by the Doppler shift of the hydrogen HI line. The observed speed is faster than one which is expected from the visible mass distribution of the galaxy. This observation indicates that there is a significant amount of unknown mass called dark matter in the galaxy. From [4]

References

1. Jackson, J.D.: Classical Electrodynamics. Wiley, Hoboken (1967)
2. Griffiths, D.J.: Am. J. Phys. **50**, 698 (1982)
3. Greiner, W., Reinhardt, J.: Quantum Electrodynamics, 2nd edn. Springer, Berlin (1994)
4. Siegel, E.: Forbes magazine, 29 June 2017

Anomalous Magnetic Moment

4

4.1   Introduction

The magnetic moment of spin-1/2 fermion with mass m and charge e is predicted as $\mu = e/2m$ from the Dirac equation. The fermion precesses in magnetic field (B) with angular velocity $\omega_p = eB/m$ due to the magnetic moment. On the other hand, the cyclotron frequency of the fermion moving in the same magnetic field is $\omega_c = eB/m$. Since $\omega_c = \omega_p$, the helicity (spin component of the momentum direction) of the fermion conserves in any magnetic field. However, it is observed that there is actually a slight difference between ω_c and ω_p. The origin of the difference is that the higher order diagrams add very small spin-flipping transition amplitude to the original one and change ω_p. This effect is important because we can look new physics out through the additional transition amplitude.

4.2   Helicity Conservation

In a homogeneous magnetic field $\vec{B} = (0, 0, -B)$, a particle with mass m and charge e circulates a projected round orbit to the xy-plane with cyclotron frequency $f_c = 2\pi/\omega_c$ where ω_c is the angular velocity;

$$\omega_c = \frac{e}{m}B . \tag{4.1}$$

This means that the charged particle changes its direction of motion with this rate as shown in Fig. 4.1a. On the other hand, as we learned in Chap. 2, the angular velocity of the precession of a spin-1/2 fermion is $\omega_p = 2\mu B$. For Dirac fermion, the magnetic moment is predicted as $\mu = e/2m$ and the angular velocity of the spin precession

© Springer Nature Switzerland AG 2021
F. Suekane, *Quantum Oscillations*, Lecture Notes in Physics 985,
https://doi.org/10.1007/978-3-030-70527-5_4

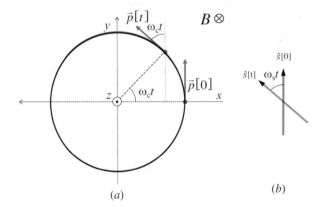

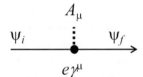

Fig. 4.1 **a** Cyclotron revolution and **b** spin precession in a magnetic field B. For $g = 2$, the angular velocity of the rotation of the momentum direction and the spin direction are the same and the helicity is conserved

Fig. 4.2 The vertex of electromagnetic interaction of fermion

becomes, from Eq. (2.38),

$$\omega_p = \frac{e}{m} B \ . \tag{4.2}$$

Therefore, the angular velocity of the spin precession (4.2) coincides with the angular velocity of the cyclotron frequency (4.1),

$$\omega_c = \omega_p = \frac{e}{m} B \ . \tag{4.3}$$

This means that the relation of the spin direction and the direction of motion do not change and the helicity of fermions is not affected by the magnetic field even if the trajectory is deflected. For the relativistic situation, both angular velocities slow down by the Lorentz factor due to the time delation but the relation between both directions holds. This phenomenon is called the helicity conservation.

This coincidence comes from the structure of the electromagnetic interactions shown in Fig. 4.2. The amplitude of the electromagnetic interaction is

$$\mathcal{M}_{fiA} = -e \left\{ \overline{\psi}_f \gamma^\mu \psi_i \right\} A_\mu \ . \tag{4.4}$$

If we take the low energy limit of Eq. (4.4), we obtain the amplitude of the Schrödinger-Pauli equation [1],

$$\mathcal{M}_{fiA} \to -e \left(A^0 - \frac{1}{m} (\vec{P} \cdot \vec{A}) \right) \{ \psi_f^\dagger \psi_i \} + \frac{e}{2m} \left(\vec{B} \cdot \{ \psi_f^\dagger \vec{\sigma} \psi_i \} \right) \ , \tag{4.5}$$

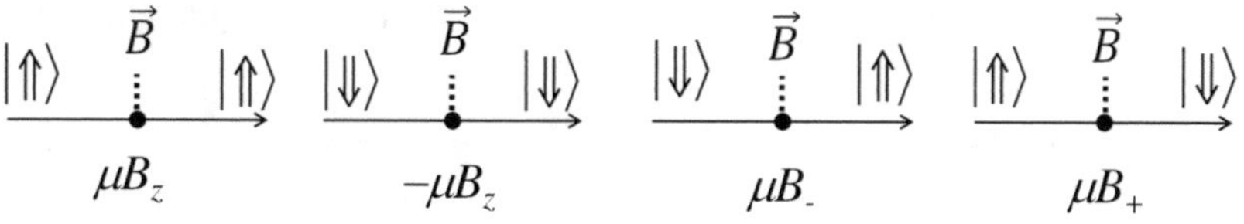

$$|\Uparrow\rangle \quad \overset{\vec{B}}{\vdots} \quad |\Uparrow\rangle \qquad |\Downarrow\rangle \quad \overset{\vec{B}}{\vdots} \quad |\Downarrow\rangle \qquad |\Downarrow\rangle \quad \overset{\vec{B}}{\vdots} \quad |\Uparrow\rangle \qquad |\Uparrow\rangle \quad \overset{\vec{B}}{\vdots} \quad |\Downarrow\rangle$$

$$\mu B_z \qquad\qquad -\mu B_z \qquad\qquad \mu B_- \qquad\qquad \mu B_+$$

Fig. 4.3 Spin transition amplitudes

where $\vec{P}$ is the momentum of the particle and the relation $\vec{B} = (\vec{\nabla} \times \vec{A})$ is used. The two terms in the first parenthesis cause the deflection of the particle trajectory due to the electromagnetic field. The A_0 term corresponds to the static electric field, which we assume to be 0 in this chapter. The second term causes the deflection of the trajectory by the magnetic field $\vec{B}$. These terms do not change the spin direction. The third term represents the interaction between the magnetic field $\vec{B}$ and magnetic moment; $\vec{\mu} = e/2m\vec{\sigma}$ which is equivalent to Eq. (2.4). This term causes the spin precession and changes the spin direction. The wave function can be expressed as a superposition of spin-up $(|\Uparrow\rangle)$ and spin-down $(|\Downarrow\rangle)$ states as

$$\psi_i = \begin{pmatrix} \alpha_i \\ \beta_i \end{pmatrix} = \alpha_i |\Uparrow\rangle + \beta_i |\Downarrow\rangle, \quad \psi_f = \begin{pmatrix} \alpha_f \\ \beta_f \end{pmatrix} = \alpha_f |\Uparrow\rangle + \beta_f |\Downarrow\rangle, \qquad (4.6)$$

and the third term of Eq. (4.5) can be expanded as

$$M_{\mu B} = \mu \begin{pmatrix} \alpha_f^* & \beta_f^* \end{pmatrix} \begin{pmatrix} B_z & B_- \\ B_+ & -B_z \end{pmatrix} \begin{pmatrix} \alpha_i \\ \beta_i \end{pmatrix} = \mu \left(\alpha_f^* \alpha_i B_z + \alpha_f^* \beta_i B_- + \beta_f^* \alpha_i B_+ - \beta_f^* \beta_i B_z \right) .$$
$$(4.7)$$

The transition amplitude for $|\Uparrow\rangle \rightarrow |\Downarrow\rangle$ corresponds to the case $(\alpha_i, \beta_i) = (1, 0)$ and $(\alpha_f, \beta_f) = (0, 1)$ and $M_{\mu B} = \mu B_+$. Figure 4.3 shows the spin transitions corresponding to Eq. (4.7) and the corresponding Pauli equation becomes

$$i\frac{d}{dt}\psi = -\mu(\vec{B} \cdot \vec{\sigma})\psi = -\frac{1}{2}\omega_p(\hat{B} \cdot \vec{\sigma})\psi , \qquad (4.8)$$

where $\hat{B}$ is a unit vector with direction $(\vec{B})$. This equation is the same as Eq. (2.4) after replacing

$$\omega_B \rightarrow -\frac{\omega_p}{2} \qquad (4.9)$$

and we can borrow the general wave function (2.65) obtained there.

The vector of the magnetic field shown in Fig. 4.1 is $\vec{B} = (0, 0, -B)$. The corresponding polar coordinates are $(\theta, \phi) = (\pi, 0)$. Therefore, the general spin wave function of our case is, from Eq. (2.65),

$$\Psi_B[t] = \alpha[0]e^{-i(\omega_p/2)t}|\Uparrow\rangle + \beta[0]e^{i(\omega_p/2)t}|\Downarrow\rangle . \qquad (4.10)$$

From Eq. (2.47), the wave function of the spin that is pointing to $+y$ direction is

$$|\otimes\rangle = \frac{1}{\sqrt{2}}(|\Uparrow\rangle + i|\Downarrow\rangle) . \qquad (4.11)$$

If the spin is pointing to $+y$ direction at time $t = 0$, the initial condition is

$$\Psi_B[t = 0] = \alpha[0]|\uparrow\rangle + \beta[0]|\downarrow\rangle = |\otimes\rangle = \frac{|\uparrow\rangle + i|\downarrow\rangle}{\sqrt{2}} . \tag{4.12}$$

Therefore, $(\alpha[0], \beta[0]) = (1/\sqrt{2}, i/\sqrt{2})$. Then the wave function at time t becomes

$$\Psi_B[t] = e^{(\pi/4)i} \begin{pmatrix} e^{-i(\omega_p t + \pi)/4} \cos[\pi/4] \\ e^{i(\omega_p t + \pi)/4} \sin[\pi/4] \end{pmatrix} . \tag{4.13}$$

This is the state that the spin direction is $(\theta, \phi) = (\pi/2, (\omega_p/2)t + \pi/2)$. The spin direction is precessing in the xy-plane with angular velocity ω_p as expected.

4.2.1 Anomalous Magnetic Moment

The amplitude of the actual interaction is the sum of all the effects including higher order corrections as shown in Fig. 4.4. Due to the vertex correction shown in Fig. 4.4b, the spin transition part of the vector current changes as[1]

$$\mathcal{M} = -\mu\vec{B} \cdot \{u_f^\dagger \vec{\sigma} u_i\} \rightarrow -\left(1 + \frac{\alpha}{2\pi}\right) \mu\vec{B} \cdot \{u_f^\dagger \vec{\sigma} u_i\} , \tag{4.14}$$

where the first term comes from Fig. 4.4a and the second term comes from Fig. 4.4b. This means we can regard that the magnetic moment changes as

$$\mu \rightarrow \mu' = \left(1 + \frac{\alpha}{2m}\right)\mu \tag{4.15}$$

due to the vertex correction. For $\mu^\pm$ particles, the contributions from higher order diagrams (Fig. 4.5) are

$$a_\mu^{SM} \equiv \frac{\mu' - \mu}{\mu} = \frac{1}{2}\left(\frac{\alpha}{\pi}\right) + 0.7568574\left(\frac{\alpha}{\pi}\right)^2 + \cdots$$
$$= 0.00116591823 \pm 0.00000000043 . \tag{4.16}$$

In quantum mechanics, traditionally we relate the magnetic moment μ and angular momentum L using g-factor, defined in

$$\mu = g\frac{e}{2m}L . \tag{4.17}$$

---

[1] For example, Eq. (7.37) of [1].

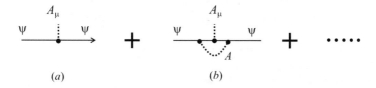

Fig. 4.4 Higher order contributions. **a** Tree diagram. **b** Vertex correction

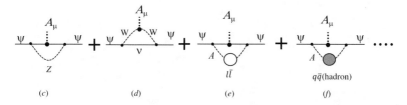

Fig. 4.5 Additional vertex corrections. **c, d** Weak interactions. **e** Higher order QED, **f** Strong interactions

For the magnetic moment of the orbital angular momentum, $g = 1$. However, for the magnetic moment of the spin of Dirac fermions, the angular momentum is $L = S = 1/2$ but the magnetic moment is still $\mu = \frac{e}{2m}$. Therefore,

$$\mu = 2 \frac{e}{2m} S \qquad (4.18)$$

and the g factor is 2. Deviation from $g = 2$ is called *anomalous magnetic moment* and we often express it as $g - 2$ (pronounced as "gee minus two"). For electron and muon, the $g - 2$ value can be calculated accurately and can also be measured accurately and strict test of the theory can be performed by using $g - 2$ values. Historically the correctness of the quantum electro dynamics (QED) was confirmed from the agreement between predictions and measurements of the $g - 2$ values. If significant deviation is found between the prediction and the measurement, it is a hint of something we have not known yet exists in the nature.

4.2.2 Measurements

4.2.2.1 Muon $g - 2$

$g - 2$ of muon was measured accurately as follows. High energy proton beam hits a target and produces π^+. The π^+ decays in flight and produces μ^+ and ν_μ.

$$\pi^+ \to \mu^+ + \nu_\mu . \qquad (4.19)$$

This decay takes place through the weak interaction and the helicity of ν_μ is 100% negative in the π^+ rest frame. Since π^+ spin is 0, the helicity of μ^+ is also 100% negative as shown in Fig. 4.6.

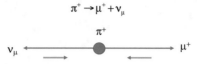

$$\pi^+ \to \mu^+ + \nu_\mu$$

Fig. 4.6 π^+ decay. Helicity of μ^+ is 100% negative

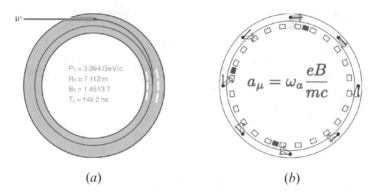

(a)   (b)

Fig. 4.7 a $g-2$ muon storage ring. P_0 is the "magic" muon momentum that cancels the effect of the electric field, R_0 is the orbital radius, B_0 is the magnetic field strength of the storage ring magnets, and T_C is the cyclotron period. **b** Polarimeter. The line arrows are the muon momentum, the open arrows are the direction of muon spin, corresponding to the positron favored direction. Open boxes are calorimeters to detect e^+. From [2]

Fig. 4.8 μ^+ decay and "right combination" of the spins. High energy e^+ tends to be emitted along the μ^+ spin direction

$$\mu^+ \to e^+ + \nu_e + \overline{\nu}_\mu$$

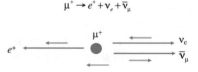

This polarized μ^+'s are sent to a storage ring (Fig. 4.7a) and they circulate in the ring orbit and eventually decay

$$\mu^+ \to e^+ + \nu_e + \overline{\nu}_\mu . \tag{4.20}$$

This decay is caused by the weak interaction and ν_e has negative helicity, $\overline{\nu}_\mu$ has positive helicity and high energy e^+ tends to have positive helicity.

In addition, the total spin of e^+ and ν_e is $J = 1$, since they are produced by the decay of the spin-1 W^+ boson. Therefore, e^+ tends to be emitted along the μ^+ spin as shown in Fig. 4.8. This means the μ^+ polarization can be measured from the angular distribution of e^+ emission.

Positron detectors are arranged along the inner wall of the storage ring as shown in Fig. 4.7b and the e^+ emission angle is measured. If $(g_\mu - 2)/2 = a_\mu$ is not zero, the muon spin rotates with respect to its direction of motion with an angular velocity

$$\omega_a = \omega'_p - \omega_p = a_\mu \frac{e}{m} B . \tag{4.21}$$

Fig. 4.9 60 h data set.
Horizontal axis is time and
vertical axis is the count rate
of the positron. From [2]

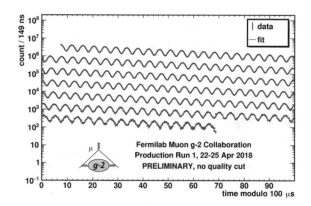

Figure 4.9 shows the time dependence of the count rate of the positron counters. The
overall decreasing property is due to the decay and loss of μ^+ in the storage ring.
The small ripples correspond to the spin rotation (ω_a) due to the anomalous magnetic
moment. If g is exactly 2, the spin direction is always antiparallel to the direction of
the movement and the count rate would not change.

Finally, the most accurate $g - 2$ measurement to date is [3]

$$a_\mu^{\text{exp}} = 0.0011659209 \pm 0.0000000006 . \qquad (4.22)$$

If we compare this value with the prediction, Eq. (4.16),

$$\frac{a_\mu^{\text{exp}}}{a_\mu^{\text{SM}}} = 1.00000229 \pm 0.00000052 . \qquad (4.23)$$

They agree with each other with very good precision. However, there is $\sim 4\sigma$ discrepancy from the perfect agreement. Currently it is not clear whether the discrepancy
comes from some artifact of experiment and/or theory or if it is a hint of new physics.
For the theoretical point of view, it is difficult to calculate the contribution from the
strong interaction (Fig. 4.5d). For the experimental point of view, for example, it is
difficult to guarantee that there is no electrostatic effect.

Electron anomalous magnetic moment a_e is calculated and measured with an even
better precision than a_μ,

$$
\begin{aligned}
a_e^{\text{exp}} &= 0.00115965218091 \pm 0.00000000000026 \ (\text{PDG2018}), \\
a_e^{\text{SM}} &= 0.0011596521818 \ \pm \ 0.0000000000076 \ (\text{arXiv} : 1208.6583\text{v2}),
\end{aligned}
\qquad (4.24)
$$

and

$$\frac{a_e^{\text{exp}}}{a_e^{\text{SM}}} = (1 - 0.00000000077) \pm 0.00000000066 \,. \qquad (4.25)$$

They agree within $\sim 1\sigma$.

References

1. Halzen, F., Martin, A.D.: Quarks and Leptons. Wiley, Hoboken (1984)
2. Keshavarzi, A.: Muon g-2 collaboration. arXiv:1905.00497v2 [hep-ex] 3 May 2019
3. Particle Data Group: Prog. Theor. Exp. Phys. **2020**, 083C01 (2020)

Positronium

5

5.1   Introduction

The positronium, often labeled as "Ps", is a e^+-e^- binding state as shown in Fig. 5.1. In the positronium, electron and positron are interacting with each other by the electromagnetic force,

$$e^- + e^+ \rightarrow e^- + e^+. \tag{5.1}$$

Since the system includes only elementary particles and electromagnetic interactions, its behavior is very well understood quantitatively by the quantum electrodynamics (QED). In addition, the positronium system includes various transitions and oscillation effects and is a complex superposition states. Therefore, the positronium is a very good example to understand such effects.

The energy level structure of the positronium is similar to that of the hydrogen atom. However, there is a difference due to e^+e^- annihilation process.

The QED diagrams of the positronium are shown in Fig. 5.2. The T-channel process takes place due to the emission and absorption of the virtual photon. This process generates the electrostatic potential and magnetic moment-magnetic moment (MM-MM) interaction, like the hydrogen atom. The S-channel process is unique to the positronium in which e^+-e^- annihilates into virtual photon and the virtual photon creates a e^+-e^- pair. This effect changes the energy level structure.

Fig. 5.1 Positronium (Ps) is a QED binding state of e^+ and e^-

© Springer Nature Switzerland AG 2021
F. Suekane, *Quantum Oscillations*, Lecture Notes in Physics 985,
https://doi.org/10.1007/978-3-030-70527-5_5

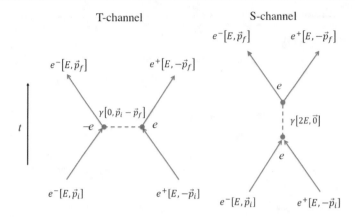

Fig. 5.2 Diagrams of $e^+e^- \rightarrow e^+e^-$ interactions. The scattering process is called T-channel and the annihilation process is called S-channel interactions

Taking the low energy limits, the Hamiltonian of the positronium system with angular momentum $l = 0$ state is expressed as[1]

$$H_{Ps} \sim H_K + H_M + H_A \,, \tag{5.2}$$

where H_K is kinetic and electrostatic terms,

$$H_K = \frac{p^2}{m_e} - \frac{\alpha}{r} \,. \tag{5.3}$$

The p^2/m_e term corresponds to the kinetic energy of $e^\pm$ system which has the reduced mass, $\mu = m_e/2$.

H_M is for MM-MM interaction at low momentum limit of the T-channel diagram of Fig. 5.2,

$$H_M = \frac{2\pi\alpha}{3m_e^2}(\vec{\sigma}_+ \cdot \vec{\sigma}_-)\delta^3[\vec{r}] \,, \tag{5.4}$$

where the Pauli spin vector $\vec{\sigma}_\pm$ represents the unit spin vector of $e^\pm$ ($\vec{s} = \frac{1}{2}\vec{\sigma}$), respectively.

H_A is for pair annihilation-creation process in the S-channel diagram,

$$H_A = \frac{\pi\alpha}{2m_e^2}(3 + (\vec{\sigma}_+ \cdot \vec{\sigma}_-))\delta^3[\vec{r}] \,. \tag{5.5}$$

---

[1] Example 6.5 of [1].

5.2 H_K: e^+-e^- **Binding State by Electrostatic Potential**

The H_K shows a motion of a charged particle with charge $-e$ and reduced mass μ in the electrostatic potential $V[r] = \frac{e}{4\pi r}$. The corresponding Schrödinger equation is

$$i\frac{d}{dt}\psi_K[t,\vec{r}] = \left(-\frac{1}{2\mu}\vec{\nabla}^2 - \frac{\alpha}{r}\right)\psi_K[t,\vec{r}]. \tag{5.6}$$

This is exactly the same form as the Schrödinger equation for the hydrogen atom. The only difference is the reduced mass (for hydrogen atom, $\mu = m_p m_e/(m_p + m_e) \sim m_e$) of the equation. Therefore, the space structure of the positronium is the same as that of the hydrogen except for that the scale of space and energy is two times different. We can adopt various properties of the hydrogen atom quantitatively by just replacing $m_e \rightarrow \mu = m_e/2$. For example, the average radius (a_{Ps}) and energy of the $1S$ state of the positronium (Ps) can be obtained from those of the hydrogen atom as

$$a_H = a_B = \frac{1}{m_e\alpha} = 0.53 \times 10^{-10}\,\text{m} \quad \rightarrow \quad a_{Ps} = \frac{2}{m_e\alpha} = 1.06 \times 10^{-10}\,\text{m},$$

$$E_H = -\frac{m_e\alpha^2}{2} = -13.6\,\text{eV} \quad \rightarrow \quad E_{Ps} = -\frac{m_e\alpha^2}{4} = -6.8\,\text{eV}. \tag{5.7}$$

The space-time part of the wave function of the positronium is

$$\psi_K^{nlm}(t,\vec{r}) = R_{nl}[r]Y_l^m[\theta,\phi]e^{-iE_n t}, \tag{5.8}$$

where Y_l^m is the spherical harmonics for orbital angular momentum l and its z-component m and $|R_{nl}[r]|^2$ shows r dependence of the probability density function for the principal quantum number n that represents to the radial excitation. The energy level depends only on the principal quantum number n. In this text, we only deal with the ground state; $n = 1$ and thus $l = 0$ and $m = 0$. The kinetic part of the wave function is, then

$$\psi_K^0[t,\vec{x}] = \frac{1}{\sqrt{\pi a_{Ps}^3}}\exp\left[-\frac{r}{a_{Ps}}\right]\exp[-iE_{Ps}t]. \tag{5.9}$$

5.2.1 H_M: **MM-MM Interactions**

The actual effect of the Hamiltonian, $H[\vec{x}]$ is the weighted average of H with the weight of the probability function of the ψ_K,

$$\mathcal{H}_M = \int H_M[\vec{x}]\,|\psi_K[\vec{x}]|^2\,d^3\vec{x} = \frac{\alpha^4 m_e}{12}(\vec{\sigma}_+ \cdot \vec{\sigma}_-) \equiv A_M(\vec{\sigma}_+ \cdot \vec{\sigma}_-), \tag{5.10}$$

where $A_M = \frac{\alpha^4}{12} m_e \sim 120.8\,\mu eV$ is the scale of the energy split. The corresponding equation of motion for the MM-MM interaction is then

$$i\frac{d}{dt}\psi_M = A_M(\vec{\sigma}_+ \cdot \vec{\sigma}_-)\psi_M \;. \tag{5.11}$$

This is the same form as the hydrogen hyperfine splitting case shown in Eq. (3.11) and the discussions there can directly be adopted here. From the analogy to the discussions there, the general spin-part wave function is a superposition of the $e^{\pm}$ spin states,

$$\psi_M[t] = C^M_{\Uparrow\Uparrow}[t]|\Uparrow_{e^+}\Uparrow_{e^-}\rangle + C^M_{\Uparrow\Downarrow}[t]|\Uparrow\Downarrow\rangle + C^M_{\Downarrow\Uparrow}[t]|\Downarrow\Uparrow\rangle + C^M_{\Downarrow\Downarrow}[t]|\Downarrow\Downarrow\rangle. \tag{5.12}$$

The coefficients propagate in time based on the following differential equation:

$$i\frac{d}{dt}\begin{pmatrix} C^M_{\Uparrow\Uparrow} \\ C^M_{\Uparrow\Downarrow} \\ C^M_{\Downarrow\Uparrow} \\ C^M_{\Downarrow\Downarrow} \end{pmatrix} = A_M \begin{pmatrix} 1 & 0 & 0 & 0 \\ 0 & -1 & 2 & 0 \\ 0 & 2 & -1 & 0 \\ 0 & 0 & 0 & 1 \end{pmatrix} \begin{pmatrix} C^M_{\Uparrow\Uparrow} \\ C^M_{\Uparrow\Downarrow} \\ C^M_{\Downarrow\Uparrow} \\ C^M_{\Downarrow\Downarrow} \end{pmatrix}. \tag{5.13}$$

The energy eigenstates and energies are

$$\psi^I_M[t] = e^{-iA_M t}|\Uparrow\Uparrow\rangle, \quad \psi^{II}_M[t] = e^{-iA_M t}|\Downarrow\Downarrow\rangle, \quad \psi^{III}_M[t] = e^{-iA_M t}\frac{|\Uparrow\Downarrow\rangle + |\Downarrow\Uparrow\rangle}{\sqrt{2}},$$

$$\psi^{IV}_M[t] = e^{3iA_M t}\frac{|\Uparrow\Downarrow\rangle - |\Downarrow\Uparrow\rangle}{\sqrt{2}}\;. \tag{5.14}$$

ψ_I, ψ_{II} and ψ_{III} are spin-1 states and ψ_{IV} is spin-0 state. A similar energy level structure as shown in Fig. 3.3 for the hydrogen case is obtained here, although the size is much larger in this case due to much larger magnetic moment of the positron compared with that of the proton. The energy spread here is

$$\Delta E_M = 4A_M = \frac{\alpha^4}{3} m_e \sim 484\,\mu eV\;. \tag{5.15}$$

Note that the value of the positronium hyperfine energy split (5.15) can be derived from the HFS of the hydrogen 1S state, Eq. (3.24) as follows. The magnetic moment of the positron is 657 times larger than that of the proton due to its much smaller mass,

$$\frac{\mu_{e^+}}{\mu_p} = \frac{e}{2m_e}\frac{2m_p}{\kappa_p e} = \frac{m_p}{2.793 m_e} \sim 657\;, \tag{5.16}$$

where κ_p is the factor of the anomalous magnetic moment of the proton . On the other hand, because the average radius of the 1S orbit of the positronium is two times larger than that of the hydrogen, the density of the particle overlapping is eight times smaller for the positronium case than for the hydrogen atom,

$$\frac{|\psi_{\text{Ps}}[0]|^2}{|\psi_{\text{H}}[0]|^2} = \left(\frac{a_B}{a_{\text{Ps}}}\right)^3 = \frac{1}{8} . \tag{5.17}$$

Therefore, using the measured hydrogen hyperfine splitting shown in Eq. (3.27), the energy split due to MM-MM interaction in the positronium is expected as

$$E_{dd}^{\text{Ps}} = E_{dd}^{H} \times \frac{m_p}{8\kappa_p m_e} = 5.87433 \,\mu\text{eV} \times \frac{1}{8} \times 657 \sim 482 \,\mu\text{eV} . \tag{5.18}$$

This agrees with Eq. (5.15).

5.2.2 H_A: Pair Annihilation and Creation

The actual Hamiltonian of pair annihilation and creation is, like Eq. (5.10),

$$\mathcal{H}_A = \int H_A[\vec{r}] \, |\psi_K[x]|^2 \, d^3\vec{x} = \frac{\alpha^4 m_e}{16} (3 + (\vec{\sigma}_+ \cdot \vec{\sigma}_-)) \equiv A_A (3 + (\vec{\sigma}_+ \cdot \vec{\sigma}_-)) , \tag{5.19}$$

where $A_A = \frac{\alpha^4}{16} m_e \sim 90.7 \,\mu\text{eV}$ is the scale of the energy shift due to the pair annihilation-creation process. The corresponding equation of motion for the MM-MM interaction is then,

$$i \frac{d}{dt} \psi_A = A_A (3 + (\vec{\sigma}_+ \cdot \vec{\sigma}_-)) \psi_A . \tag{5.20}$$

The general wave function can be expressed as

$$\psi_A[t] = C_{\uparrow\uparrow}^A[t] |\Uparrow_{e^+} \Uparrow_{e^-}\rangle + C_{\uparrow\downarrow}^A[t] |\Uparrow\Downarrow\rangle + C_{\downarrow\uparrow}^A[t] |\Downarrow\Uparrow\rangle + C_{\downarrow\downarrow}^A[t] |\Downarrow\Downarrow\rangle. \tag{5.21}$$

Since the equation of motion for $\mathcal{H} \propto (\sigma_+ \cdot \sigma_-)$ is expressed as (5.13), that for $\mathcal{H} \propto (3 + (\sigma_+ \cdot \sigma_-))$ is expressed as follows:

$$i \frac{d}{dt} \begin{pmatrix} C_{\uparrow\uparrow}^A \\ C_{\uparrow\downarrow}^A \\ C_{\downarrow\uparrow}^A \\ C_{\downarrow\downarrow}^A \end{pmatrix} = 2 A_A \begin{pmatrix} 2 & 0 & 0 & 0 \\ 0 & 1 & 1 & 0 \\ 0 & 1 & 1 & 0 \\ 0 & 0 & 0 & 2 \end{pmatrix} \begin{pmatrix} C_{\uparrow\uparrow}^A \\ C_{\uparrow\downarrow}^A \\ C_{\downarrow\uparrow}^A \\ C_{\downarrow\downarrow}^A \end{pmatrix} . \tag{5.22}$$

The general solution of Eq. (5.22) is

$$C_{\uparrow\uparrow}^A[t] = a e^{-4i A_A t}, \quad C_{\downarrow\downarrow}^A[t] = b e^{-4i A_A t}, \quad \frac{C_{\uparrow\downarrow}^A[t] + C_{\downarrow\uparrow}^A[t]}{\sqrt{2}} = c e^{-4i A_A t}$$

$$\frac{C_{\uparrow\downarrow}^A[t] - C_{\downarrow\uparrow}^A[t]}{\sqrt{2}} = d,$$

$$\tag{5.23}$$

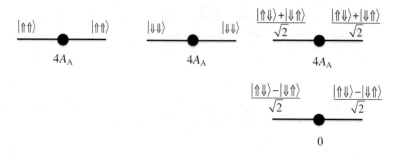

Fig. 5.3 Self-transitions due to S-channel diagram of Fig. 5.2

and from these solutions, the energy eigenstates and their energies are

$$\psi_A^I = | \Uparrow \Uparrow \rangle e^{-4iA_At}, \quad \psi_A^{II} = | \Downarrow \Downarrow \rangle e^{-4iA_At}, \quad \psi_A^{III} = \frac{| \Uparrow \Downarrow \rangle + | \Downarrow \Uparrow \rangle}{\sqrt{2}} e^{-4iA_At}$$

$$\psi_A^{IV} = \frac{| \Uparrow \Downarrow \rangle - | \Downarrow \Uparrow \rangle}{\sqrt{2}} e^{-i \times 0 \times A_At} .$$

(5.24)

Note that the energy of ψ_A^{IV} is 0. The above mass eigenstate pattern means there are self-transitions as shown in Fig. 5.3. This means that there is no effect from the annihilation diagram for the spin-0 state. This can be understood by the fact that the spin of photon is 1 and the spin-0 state cannot annihilate into the spin-1 photon.

5.2.3 $H_M + H_A$: Both Effects

So far we have seen the effects of H_M and H_A separately to obtain their physical meanings. However, in the actual world those effects cannot be separated and we have to take them into account simultaneously. In order to do so, what to do is just to add the amplitudes of the transitions as shown in Fig. 5.4.

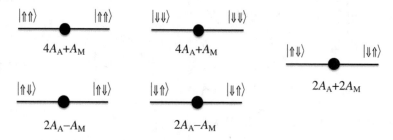

Fig. 5.4 Transition amplitudes for both MM-MM and annihilation-creation effects

The equation of motion in this case is the sum of the Hamiltonians,

$$
i\frac{d}{dt}\begin{pmatrix} C_{\uparrow\uparrow} \\ C_{\uparrow\downarrow} \\ C_{\downarrow\uparrow} \\ C_{\downarrow\downarrow} \end{pmatrix} = \begin{pmatrix} 4A_A + A_M & 0 & 0 & 0 \\ 0 & 2A_A - A_M & 2A_A + 2A_M & 0 \\ 0 & 2A_A + 2A_M & 2A_A - A_M & 0 \\ 0 & 0 & 0 & 4A_A + A_M \end{pmatrix}\begin{pmatrix} C_{\uparrow\uparrow} \\ C_{\uparrow\downarrow} \\ C_{\downarrow\uparrow} \\ C_{\downarrow\downarrow} \end{pmatrix}.
$$

$$(5.25)$$

The energy eigenstates and energies from Eq. (5.25) are

$$
\psi_I = |\uparrow\uparrow\rangle e^{-i(4A_A+A_M)t}, \qquad \psi_{II} = |\downarrow\downarrow\rangle e^{-i(4A_A+A_M)t},
$$

$$
\psi_{III} = \frac{|\uparrow\downarrow\rangle + |\downarrow\uparrow\rangle}{\sqrt{2}} e^{-i(4A_A+A_M)t}, \quad \psi_{IV} = \frac{|\uparrow\downarrow\rangle - |\downarrow\uparrow\rangle}{\sqrt{2}} e^{3iA_M t}, \qquad (5.26)
$$

as expected from Eqs. (5.14) and (5.24). This is because the wave functions of energy eigenstate have the same structure.

$\psi^{I,II,III}$ are spin-1 states and are called ortho-positronium (ψ_o) and ψ_{Ps}^{IV} is spin-0 state called para-positronium (ψ_p). Figure 5.5 shows the energy level structure of the positronium.

Finally, to complete the discussions of the positronium energy level structure, the relativistic correction has to be taken into account. The non-relativistic treatment only deals with up to the second term of the following expansion:

$$
E = \sqrt{m^2 + p^2} \sim 1 + \frac{p^2}{2m} - \frac{p^4}{8m^3} + \cdots . \qquad (5.27)
$$

The second term corresponds to the first term of Eq. (5.3). The third term of Eq. (5.27) is the relativistic correction, whose size is

$$
\Delta E_{Rel} \sim -\frac{E_{Ps}^2}{2\mu} = -\frac{\alpha^4}{16} m_e = -90\,\mu eV. \qquad (5.28)
$$

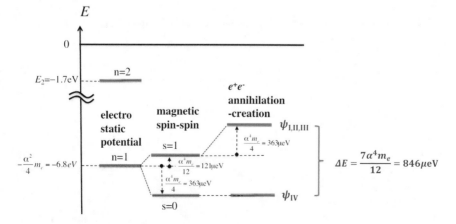

Fig. 5.5 Hyperfine energy splitting of the positronium

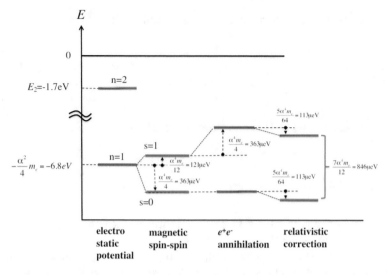

Fig. 5.6 Positronium energy level with relativistic correction

This is the similar size as A_M and A_A and cannot be ignored. A detailed calculation shows that relativistic correction is

$$\Delta E_{\text{Rel}} = -\frac{5\alpha^4}{64}m_e = -113\,\mu\text{eV}. \tag{5.29}$$

Figure 5.6 shows the energy level with the relativistic correction. Since the relativistic effect shifts the energy of both $S = 0$ and $S = 1$ states by the same amount, the energy difference does not change.

Reference

1. Greiner, W., Reinhardt, J.: Quantum Electrodynamics, 2nd edn. Springer, Berlin (1994)

Part II
Higgs Field

Weinberg Angle

6

6.1 Introduction

The Weinberg angle, or sometimes called the weak mixing angle, θ_W is one of the fundamental parameters of the standard model of the elementary particles, which have been measured accurately. Historically, strict tests of the standard model were performed using θ_W and other accurately measured parameters. The top quark mass and Higgs boson mass were predicted from calculations of the higher order diagrams using the measured θ_W and the predictions were confirmed experimentally in the end.

The Weinberg angle is a mixing angle between U(1) and SU(2) neutral gauge bosons. We recognize the two mass eigenstate after the mixing as massless photon A and heavy Z^0 particle. The U(1) and SU(2) gauge bosons oscillate in A and Z^0 but the oscillation is too quick to be observed.

In this chapter, the origin and properties of θ_W are described. At first we ignore the chirality dependence of weak interactions in order to simplify the description. Then later on, we introduce the chirality dependence to give the complete formula of the electroweak interactions based on the Weinberg angle.

6.2 General Formula of Electromagnetic and Weak Interactions

Quarks and leptons perform electromagnetic and weak interactions. Figure 6.1 shows some examples of those interactions. The electromagnetic interactions are caused by the exchange of massless photon. The weak interactions are caused by the exchange of massive weak bosons, $W^\pm$ and Z^0. Their masses are $M_W = 80.4$ GeV/c^2 and $M_Z = 91.2$ GeV/c^2, respectively. Figure 6.2 shows the fermion-boson couplings.

© Springer Nature Switzerland AG 2021
F. Suekane, *Quantum Oscillations*, Lecture Notes in Physics 985,
https://doi.org/10.1007/978-3-030-70527-5_6

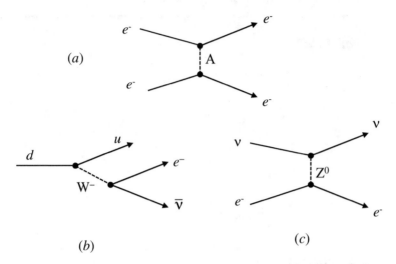

Fig. 6.1 Examples of **a** $e^- $-$e^-$ scattering by the electromagnetic interaction. **b** β decay by the weak charged current interaction. **c** ν-e^- elastic scattering by the weak neutral current interaction

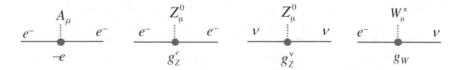

Fig. 6.2 Couplings of photons and weak bosons to electron and neutrino

The corresponding interaction Hamiltonian is

$$
\begin{aligned}
\mathcal{H}_{ffG} &= -e\{\overline{\psi_e}\gamma^\mu\psi_e\}A_\mu + 0 \times \{\overline{\psi_\nu}\gamma^\mu\psi_\nu\}A_\mu \\
&+ g_Z^e\{\overline{\psi_e}\gamma^\mu\psi_e\}Z_\mu^0 + g_Z^\nu\{\overline{\psi_\nu}\gamma^\mu\psi_\nu\}Z_\mu^0 + gw\{\overline{\psi_\nu}\gamma^\mu\psi_e\}W_\mu^+ + gw\{\overline{\psi_e}\gamma^\mu\psi_\nu\}W_\mu^- \\
&= \begin{pmatrix} \overline{\psi_\nu} & \overline{\psi_e} \end{pmatrix}\gamma^\mu \begin{pmatrix} g_Z^e Z^\mu & gw\,W_\mu^- \\ gw\,W_\mu^+ & -eA^\mu + g_Z^e Z^\mu \end{pmatrix}\begin{pmatrix} \psi_\nu \\ \psi_e \end{pmatrix}.
\end{aligned}
$$

$$(6.1)$$

In order to understand where the structure of the electroweak interactions comes from, we start the discussion with a very general boson-fermion toy interactions and try to extract Eq. (6.1) from it.

The Lagrangian density which leads the Dirac equation of electron in free space is

$$
\mathcal{L}_{ee} = \overline{\psi_e}\left(i\gamma^\mu\partial_\mu - m_e\right)\psi_e .
$$

$$(6.2)$$

In order to put the electromagnetic interaction in the Lagrangian, the derivative term ∂_μ is replaced by the so called *covariant derivative* $\mathcal{D}_\mu$,

$$
\partial_\mu \rightarrow \mathcal{D}_\mu = \partial_\mu - ieA_\mu .
$$

$$(6.3)$$

Replacing the ∂_μ by $\mathcal{D}_\mu$ in the Lagrangian (6.2), we obtain the following QED Lagrangian,

$$\mathcal{L}_{\text{QED}}^e = \overline{\psi_e}\left(\gamma^\mu(i\partial_\mu + eA_\mu) - m_e\right)\psi_e . \tag{6.4}$$

Since the neutrino charge is 0 and it is massless, its QED Lagrangian is

$$\mathcal{L}_{\text{QED}}^\nu = i\overline{\psi_\nu}\gamma^\mu\partial_\mu\psi_\nu . \tag{6.5}$$

If we combine the electron and neutrino Lagrangian into one equation, the Lagrangian is formally written as

$$\mathcal{L}_{\text{QED}}^{e\nu} = \overline{\Psi}\left(\gamma^\mu\left(i\partial_\mu + eA_\mu I_-\right) + m_e I_-\right)\Psi , \tag{6.6}$$

where

$$\Psi = \begin{pmatrix} \psi_\nu \\ \psi_e \end{pmatrix}, \quad I_- = \begin{pmatrix} 0 & 0 \\ 0 & 1 \end{pmatrix} . \tag{6.7}$$

On the other hand, the weak interactions have transitions $e \leftrightarrow \nu$ and in order to include the weak interactions in the Lagrangian, a more general interaction matrix has to be used. Due to the conservation of probability of the existence of fermions, the interaction matrix is Hermitian. Since any 2×2 Hermitian matrix can be expressed by a sum of unit matrix I and Pauli matrices $\vec{\sigma}$, the general interaction Lagrangian has the form,

$$\mathcal{L}_I = aI + (\vec{b} \cdot \vec{\sigma}) , \tag{6.8}$$

where a and $\vec{b}$ are real parameters. Therefore, the Lagrangian of the boson-(e^-, ν) interaction has a general form,[1]

$$\mathcal{L}_{ffG} = \overline{\Psi}\gamma^\mu\left(i\partial_\mu - g_0 X_\mu I - g_1(\vec{Y}_\mu \cdot \vec{\sigma})\right)\Psi , \tag{6.9}$$

where X and $\vec{Y}$ correspond to the wave function of vector bosons and g_0 and g_1 correspond to the coupling constant between the bosons and fermions. X and Y bosons are assumed to be massless at this stage.[2]

Expanding Eq. (6.9) and writing down the interaction terms explicitly, the general interaction Hamiltonian is

$$\mathcal{H}_{ffG} = \begin{pmatrix} \overline{\psi_\nu} & \overline{\psi_e} \end{pmatrix} \gamma_\mu \begin{pmatrix} g_0 X^\mu + g_1 Y_3^\mu & \sqrt{2}g_1 Y_-^\mu \\ \sqrt{2}g_1 Y_+^\mu & g_0 X^\mu - g_1 Y_3^\mu \end{pmatrix} \begin{pmatrix} \psi_\nu \\ \psi_e \end{pmatrix}$$
$$= (g_0 X_\mu - g_1 Y_{3\mu})J_{ee}^\mu + (g_0 X_\mu + g_1 Y_{3\mu})J_{\nu\nu}^\mu + \sqrt{2}g_1(Y_{-\mu}J_{e\nu}^\mu + Y_{+\mu}J_{\nu e}^\mu) , \tag{6.10}$$

where $J_{\alpha\beta}^\mu = \{\overline{\psi_\beta}\gamma^\mu\psi_\alpha\}$ and $Y_\pm^\mu = (Y_1^\mu \pm iY_2^\mu)/\sqrt{2}$. Figure 6.3 shows the boson-fermion vertices obtained from Eq. (6.10). Since $Y_\pm$ change e^- to ν and vice versa,

---

[1] In the standard model, this form is introduced to satisfy the gauge symmetry.
[2] The gauge symmetry requires so.

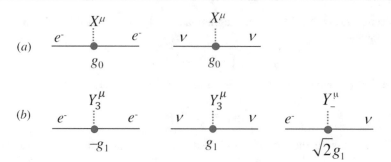

Fig. 6.3 Couplings of electron and neutrino to photons and weak bosons

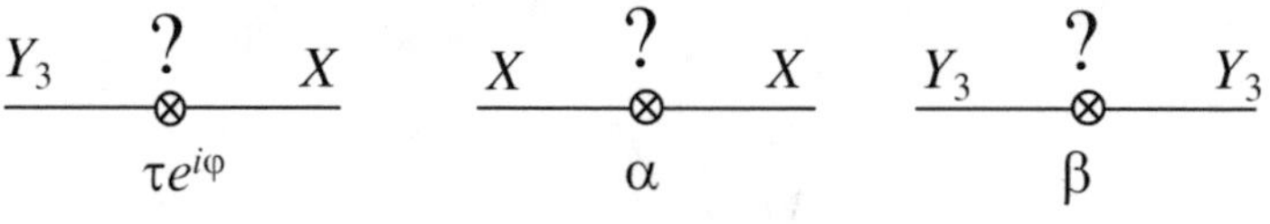

Fig. 6.4 Hypothetical transition between X and Y_3 bosons. $0 \leq \tau, 0 \leq \varphi < 2\pi, \alpha < \beta$ are assumed in the discussions. "?" shows we do not know the origin of these transitions for now

they are supposed to correspond to $W^{\mp}$ boson. X and Y_3 do not change the fermion flavor and are supposed to correspond to the photon A or Z^0. However, there is a problem. Since both X and Y_3 couple to both e^- and ν, neither X nor Y_3 can be regarded as the photon.

Here, we assume that there are transitions between X and Y_3 bosons as shown in Fig. 6.4. Due to the transitions, the general state of the neutral bosons becomes a superposition of X and Y_3,

$$|\psi_{G^0}[t]\rangle = X[t]|X\rangle + Y_3[t]|Y_3\rangle \,, \tag{6.11}$$

and the amplitudes satisfy the transition equation,

$$i\frac{d}{dt}\begin{pmatrix} Y_3 \\ X \end{pmatrix} = \begin{pmatrix} \beta & \tau e^{-i\varphi} \\ \tau e^{i\varphi} & \alpha \end{pmatrix}\begin{pmatrix} Y_3 \\ X \end{pmatrix} \,, \tag{6.12}$$

where α, β, τ and φ are real numbers. The domains of τ and φ are $0 \leq \tau$ and $0 \leq \varphi < 2\pi$, respectively. As a result, from (B.37) the mass eigenstate vectors become

$$\begin{cases} |Y_3'\rangle = \cos\theta|Y_3\rangle + e^{i\varphi}\sin\theta|X\rangle \\ |X'\rangle = -e^{-i\varphi}\sin\theta|Y_3\rangle + \cos\theta|X\rangle \,, \end{cases} \tag{6.13}$$

where θ is mixing angle defined by the mixing triangle shown in Fig. 6.5.

Fig. 6.5 Mixing triangle for
the hypothetical transitions
between X and Y_3 bosons

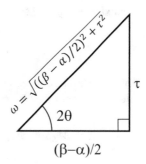

As always, the mixing angle and generated masses are expressed by using the transition amplitudes as follows:

$$\tan 2\theta = \frac{2\tau}{\beta - \alpha}, \quad \omega = \sqrt{\left(\frac{\alpha - \beta}{2}\right)^2 + \tau^2},$$

$$M_{X'} = \frac{\alpha + \beta}{2} - \omega, \quad M_{Y'} = \frac{\alpha + \beta}{2} + \omega. \tag{6.14}$$

6.2.1 Correspondence to Photon and Z^0

Equation (6.14) shows $M_{X'} < M_{Y'}$. Therefore, we would like to consider X' as the photon and Y' as Z^0. In this case, (6.13) becomes

$$\begin{cases} |Z^0\rangle = \cos\theta_W |Y_3\rangle + e^{i\varphi} \sin\theta_W |X\rangle \\ |A\rangle = -e^{-i\varphi} \sin\theta_W |Y_3\rangle + \cos\theta_W |X\rangle . \end{cases} \tag{6.15}$$

We call θ as the Weinberg angle θ_W.

Since the photon mass is 0, we require $M_{X'} = 0$,

$$M_{X'} = \frac{\alpha + \beta}{2} - \sqrt{\left(\frac{\alpha - \beta}{2}\right)^2 + \tau^2} = 0, \tag{6.16}$$

which leads to a simple relation among the transition amplitudes,

$$\underline{\alpha\beta = \tau^2}. \tag{6.17}$$

This magic relation is an important requirements for the mechanism of the X-Y_3 transition to include massless photon in the system. This relation mathematically means the determinant of the interaction matrix in Eq. (6.12) is zero and the matrix does not have the inverse. There are also simple relations,

$$\omega = \frac{\alpha + \beta}{2}, \quad \sin^2\theta_W = \frac{\alpha}{\alpha + \beta}, \quad \cos^2\theta_W = \frac{\beta}{\alpha + \beta}. \tag{6.18}$$

Next, we require $M_{Y'} = M_Z$,

$$M_{Y'} = \frac{\alpha + \beta}{2} + \omega = \alpha + \beta = M_Z . \tag{6.19}$$

Together with θ, now called θ_W, defined in Eq. (6.14), the transition amplitudes are determined from the observable parameters as follows:

$$\alpha = M_Z \sin^2 \theta_W , \quad \beta = M_Z \cos^2 \theta_W , \quad \tau = \frac{1}{2} M_Z \sin 2\theta_W . \tag{6.20}$$

Next, we express the Hamiltonian (6.10) by the physical bosons; (A, Z^0). The general wave function (6.11) expressed by $|X\rangle$ and $|Y_3\rangle$ can also be expressed by $|A\rangle$ and $|Z^0\rangle$,

$$|\psi_{G^0}[t]\rangle = X[t]|X\rangle + Y_3[t]|Y_3\rangle = A|A\rangle + Z^0|Z^0\rangle e^{-iM_Z t} . \tag{6.21}$$

Therefore, using (6.15), (X, Y_3) can be expressed by (A, Z^0) as follows:

$$\begin{aligned}
X &= A\langle X|A\rangle + Z^0\langle X|Z^0\rangle e^{-iM_Z t} = \cos\theta_W A + e^{i\varphi}\sin\theta_W Z^0 e^{-iM_Z t} , \\
Y_3 &= A\langle Y_3|A\rangle + Z^0\langle Y_3|Z^0\rangle e^{-iM_Z t} = -e^{-i\varphi}\sin\theta_W A + \cos\theta Z^0 e^{-iM_Z t} .
\end{aligned} \tag{6.22}$$

Using these relations, the (X, Y_3) part of the interaction Hamiltonian (6.10) becomes

$$\begin{aligned}
\mathcal{H}_{ffG^0} = J_{ee}^\mu &\begin{pmatrix} A_\mu(g_0\cos\theta_W + g_1 e^{-i\varphi}\sin\theta_W) \\ + Z_\mu^0(g_0 e^{i\varphi}\sin\theta_W - g_1\cos\theta_W)e^{-iM_Z t} \end{pmatrix} \\
+ J_{\nu\nu}^\mu &\begin{pmatrix} A_\mu(g_0\cos\theta_W - g_1 e^{-i\varphi}\sin\theta_W) \\ + Z_\mu^0(g_0 e^{i\varphi}\sin\theta_W + g_1\cos\theta_W)e^{-iM_Z t} \end{pmatrix} .
\end{aligned} \tag{6.23}$$

Here, $(g_0\cos\theta + g_1 e^{-i\varphi}\sin\theta)$, etc. correspond to the coupling constant between the mass eigenstate bosons and the fermion currents. Since the coupling constants between photon and electron is $-e$ and to neutrino is 0,

$$\begin{cases} g_0\cos\theta_W + g_1 e^{-i\varphi}\sin\theta_W = -e \\ g_0\cos\theta_W - g_1 e^{-i\varphi}\sin\theta_W = 0 . \end{cases} \tag{6.24}$$

From these requirements, g_0 and g_1 can be related to the electric charge as follows:

$$g_0 = -\frac{e}{2\cos\theta_W} , \quad g_1 = -\frac{e}{2\sin\theta_W} e^{i\varphi} . \tag{6.25}$$

If we require g_1 to be a real number, $e^{i\varphi}$ has to be real; $\varphi = 0$ or π. By putting the relation (6.25) into (6.23), the Hamiltonian becomes

$$\mathcal{H}_{ffG^0} = -e J_{ee} A_\mu - e^{i\varphi} \frac{e}{\sin 2\theta_W} (J_{\nu\nu} - \cos 2\theta_W J_{ee}) Z_\mu e^{-iM_Z t} . \tag{6.26}$$

By comparing with Eq. (6.1), the coupling constants are identified as

$$g_Z^e = e \cot 2\theta_W e^{i\varphi}, \quad g_Z^\nu = -\frac{e}{\sin 2\theta_W} e^{i\varphi} . \tag{6.27}$$

The $Y_\pm$ part of the interaction matrix (6.10) is, using (6.25),

$$\mathcal{H}_{ffY_\pm} = \frac{e}{\sqrt{2}\sin\theta_W} \left(J_{e\nu}^\mu Y_-^\mu + J_{\nu e}^\mu Y_+^\mu \right) . \tag{6.28}$$

By comparing with (6.1), we can regard $Y_\pm$ as $W^\pm$ and the coupling constant to fermion is

$$g_W = -\frac{e}{\sqrt{2}\sin\theta_W} e^{i\varphi} . \tag{6.29}$$

In summary for this section, we first introduced massless vector bosons (X, Y) in the (e^-, ν) system as a general Lagrangian (6.9). Then we assumed there are hypothetical self- and cross-transitions between X and Y as shown in Fig. 6.4. Due to the transitions, superpositions of the (X, Y) bosons become mass eigenstates that can be identified as physical bosons (A, Z^0) as shown in Eq. (6.15). The Weinberg angle (θ_W) is the mixing angle of the mass eigenstates. From the fact that the photon mass is 0, a relation (6.17) among the transition amplitudes are derived and each transition amplitude is related to the observable parameters, M_Z and θ_W as shown in Eq. (6.20). Next, from the charges of the electron and neutrino, the coupling constants of the weak bosons g_Z^e, g_Z^ν and g_W are related to e and θ_W as in Eqs. (6.27) and (6.29). The introduction of the transitions (Fig. 6.4) connects the general interaction Hamiltonian (6.10) to the real interaction formulas, (6.26) and (6.28).

The next thing to do is to consider a natural way to introduce the vector boson transitions whose amplitudes have the relation (6.17).

6.3   The Origin of the Vector Boson Transitions; Higgs Field

In the standard model of elementary particles, it is assumed that there is a two-component scalar boson, called the Higgs boson,

$$\Phi = \begin{pmatrix} \phi^+ \\ \phi^0 \end{pmatrix} . \tag{6.30}$$

The Lagrangian density of the Higgs boson is

$$\mathcal{L}_\Phi = (\partial_\mu \Phi)^\dagger (\partial^\mu \Phi) - V(\Phi) , \tag{6.31}$$

where

$$V(\Phi) = a|\Phi|^2 + b|\Phi|^4; \quad (a < 0, \ b > 0) . \tag{6.32}$$

Just like the fermion-vector boson interactions (6.9), the two-component Higgs boson-vector bosons interactions can be introduced to the Lagrangian (6.31) by replacing

$$\partial_\mu \to \mathcal{D}_\mu = \partial_\mu + i\frac{g'}{2}X_\mu I + i\frac{g}{2}(\vec{Y}_\mu \cdot \vec{\sigma}) \,. \tag{6.33}$$

The interaction part of the Lagrangian is then,

$$\mathcal{L}_{\Phi G} = -\frac{1}{4}\left\{\Phi^\dagger\left(g'X^\mu I + g(\vec{Y}^\mu \cdot \vec{\sigma})\right)^\dagger\left(g'X_\mu I + g(\vec{Y}_\mu \cdot \vec{\sigma})\right)\Phi\right\} \,. \tag{6.34}$$

The Higgs field generates a peculiar potential and the lowest energy state occurs not at $\Phi = 0$ but at

$$\Phi = \pm\frac{1}{\sqrt{2}}\begin{pmatrix} 0 \\ v_0 \end{pmatrix} \,. \tag{6.35}$$

It means that the universe is more stable if the Higgs field exists than not exist and that the vacuum is filled with the Higgs field. v_0 is called vacuum expectation value of the Higgs field. At the bottom of the Higgs potential, the Lagrangian (6.34) becomes

$$\mathcal{L}_{\Phi G} \to -\frac{v_0^2}{8}\left(g^2 Y^2 + g'^2 X^2 - 2gg' Y_3 X\right) \,. \tag{6.36}$$

By applying the Euler-Lagrange equation to this Lagrangian density in terms of X and Y_3, we obtain the Klein-Gordon equation for X and Y_3,

$$\frac{d^2}{dt^2}\begin{pmatrix} Y_3 \\ X \end{pmatrix} = \frac{v_0^2}{4}\begin{pmatrix} -g^2 & gg' \\ gg' & -g'^2 \end{pmatrix}\begin{pmatrix} Y_3 \\ X \end{pmatrix} \,. \tag{6.37}$$

The double differential equation (6.37) can effectively be reduced to the two times of single differential equation,

$$\frac{d}{dt}\begin{pmatrix} Y_3 \\ X \end{pmatrix} = -i\frac{v_0}{2\sqrt{g^2 + g'^2}}\begin{pmatrix} g^2 & -gg' \\ -gg' & g'^2 \end{pmatrix}\begin{pmatrix} Y_3 \\ X \end{pmatrix} \,, \tag{6.38}$$

where we took only $+v_0$ without loosing the generality.[3] This corresponds to the transition equation for the neutral gauge bosons introduced as Eq. (6.12). Figure 6.6 shows the transition amplitudes. By comparing Figs. 6.4 and 6.6, "?" in Fig. 6.4 can be regarded as the Higgs field and we obtain the relations,

$$\alpha = \frac{g'^2}{2\sqrt{g^2 + g'^2}}v_0, \quad \beta = \frac{g^2}{2\sqrt{g^2 + g'^2}}v_0, \quad \tau = \frac{gg'}{2\sqrt{g^2 + g'^2}}v_0, \quad \varphi = \pi \,. \tag{6.39}$$

---

[3]This is called spontaneous symmetry breaking.

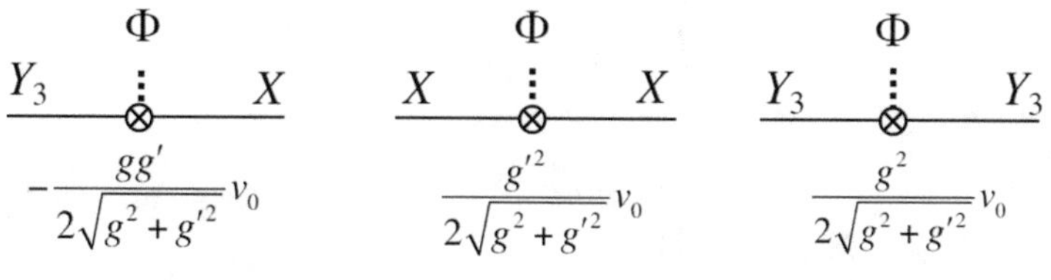

Fig. 6.6 Transition amplitudes between X and Y_3 by the Higgs field

We assumed $gg' \geq 0$ and attributed the minus sign to the phase φ. These parameters satisfy the magic relation (6.17) naturally. Therefore, the hypothesis of the Higgs mechanism satisfies the relation required in the previous section. The relation (6.17) in this case came from

$$\alpha\beta = \frac{g^2 g'^2}{4(g^2 + g'^2)} = \tau^2 . \tag{6.40}$$

This structure of the matrix can be tracked down to the binomial expansion of the Lagrangian for the general boson Lagrangian (6.34).

The physical photon and Z^0 states are, from (6.15) and $\varphi = \pi$,

$$\begin{cases} |Z^0\rangle = \cos\theta_W |Y_3\rangle - \sin\theta_W |X\rangle \\ |A\rangle = \sin\theta_W |Y_3\rangle + \cos\theta_W |X\rangle , \end{cases} \tag{6.41}$$

where the Weinberg angle θ_W is defined by the mixing triangle, Fig. 6.7a. And the Hamiltonian (6.26) is determined as

$$\mathcal{H}_{ffG^0} = -e J_{ee} A_\mu + \frac{e}{\sin 2\theta_W}(J_{\nu\nu} - \cos 2\theta_W J_{ee}) Z_\mu e^{-iM_Zt} . \tag{6.42}$$

From the mixing triangle Fig. 6.7a,

$$\tan 2\theta_W = \frac{2gg'}{g^2 - g'^2} \tag{6.43}$$

is obtained but there is a simpler expression

$$\tan\theta_W = \frac{g'}{g} \tag{6.44}$$

which is depicted in Fig. 6.7b. This expression is often used. The Z^0 mass is expressed as

$$M_Z = \frac{\alpha + \beta}{2} + \sqrt{\left(\frac{\alpha - \beta}{2}\right)^2 + \tau^2} = \frac{v_0\sqrt{g^2 + g'^2}}{2} . \tag{6.45}$$

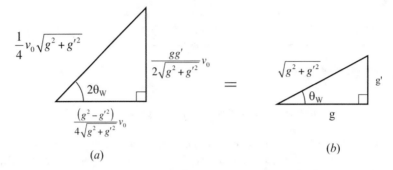

Fig. 6.7 a Mixing triangle for Higgs-vector boson coupling. **b** A Simpler expression of the Weinberg angle θ_W

As for the $W^\pm$ boson, the $W^\pm$ part of the Lagrangian (6.36) is

$$-\mathcal{L}_{\Phi Y^\pm} = \frac{v_0^2 g^2}{8} \left(Y_1^2 + Y_2^2 \right) = \frac{v_0^2 g^2}{2} \left(Y_+^* Y_+ + Y_-^* Y_- \right)$$

$$= \frac{v_0^2 g^2}{2} \left(W^{+*} W^+ + W^{-*} W^- \right) . \tag{6.46}$$

Therefore, the mass of $W^\pm$ is

$$M_W = \frac{g}{\sqrt{2}} v_0 \quad . \tag{6.47}$$

The $W^\pm$-fermion coupling constant (6.29) can be fixed by using the relation $\varphi = \pi$ as

$$g_W = \frac{e}{\sqrt{2} \sin \theta_W} \quad . \tag{6.48}$$

The fact that the mass eigenstate Z^0 with mass M_Z and A with mass $M_A = 0$ are mixtures of the original gauge bosons Y_3 and X as shown in (6.41) means that Y_3 and X oscillate within Z^0 and A with oscillation probability

$$P_{B \Leftrightarrow W_3}[t] = \sin^2 2\theta_W \sin^2 \left[\frac{1}{2} M_Z t \right] . \tag{6.49}$$

The oscillation amplitude is $\sin^2 2\theta_W \sim 0.71$ and frequency is $\omega \sim 10^{26}/s$, which is too quick to observe.

6.4 Chirality Dependence of the Weak Interactions

The weak interactions depend on the chirality of the fermion,[4] ψ_L and ψ_R. Experimentally it has been observed that $W^\pm$ do not couple to e_R^- nor ν_R, photon couples

---

[4]See Chap. 7 for chirality.

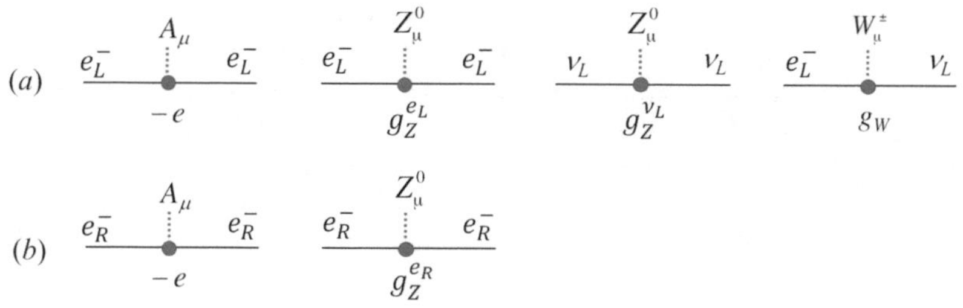

Fig. 6.8 Couplings of photons and weak bosons to electron and neutrino. **a** For left-handed chirality. **b** For right-handed chirality

to both e_L^- and e_R^- with the same coupling constant $-e$, Z^0 couples to both e_L^- and e_R^- but with different strength. Figure 6.8 shows the couplings between e^-, ν and vector bosons explicitly showing the chirality. The structure of left-handed fermion couplings shown in Fig. 6.8a is the same as Fig. 6.2 and we can follow the same discussions in the previous sections and adopt the conclusions by just replacing $\psi_e \to \psi_{eL}$ and $\psi_\nu \to \psi_{\nu L}$. From Eq. (6.27) and $\varphi = \pi$,

$$g_Z^{e_L} = g_Z^e = -e \cot 2\theta_W, \quad g_Z^{\nu_L} = g_Z^\nu = \frac{e}{\sin 2\theta_W} . \tag{6.50}$$

For right-handed fermions (Fig. 6.8b), only electron and neutral currents (photon and Z^0) participate and the minimum assumption is to add the following term to the Lagrangian:

$$-\mathcal{L}_{e_R e_R X} = g_3 \overline{e_R} \gamma^\mu X_\mu e_R = g_3 \{\overline{e_R} \gamma^\mu e_R\}(\cos \theta_W A_\mu - \sin \theta_W Z_\mu^0) . \tag{6.51}$$

From the electric charge,

$$g_3 \cos \theta_W = -e \to g_3 = -\frac{e}{\cos \theta_W} \tag{6.52}$$

and then the coupling of e_R^- and Z^0 is

$$g_Z^{e_R} = -g_3 \sin \theta_W = e \tan \theta_W . \tag{6.53}$$

6.5   Test of the Electroweak Theory

Table 6.1 summarizes the relations between the physics parameters and the electroweak parameters. Combinations of these parameters can be measured and strict tests were performed by experimental data and now we believe the electroweak theory is correct.

Table 6.1 Relation between physics parameters and electroweak parameters

M_A	M_W	M_Z	$\tan\theta_W$	e	g_W	$g_Z^{e_L}$	$g_Z^{e_R}$	$g_Z^{v_L}$
0	$\dfrac{g}{\sqrt{2}}v_0$	$\dfrac{v_0\sqrt{g^2+g'^2}}{2}$	$\dfrac{g'}{g}$	$\dfrac{gg'}{g^2+g'^2}$	$\dfrac{e}{\sqrt{2}\sin\theta_W}$	$-e\cot 2\theta_W$	$e\tan\theta_W$	$\dfrac{e}{\sin 2\theta_W}$

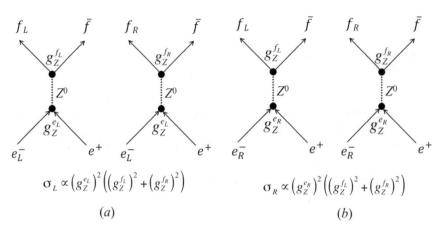

$$\sigma_L \propto \left(g_Z^{e_L}\right)^2\left(\left(g_Z^{f_L}\right)^2+\left(g_Z^{f_R}\right)^2\right) \qquad \sigma_R \propto \left(g_Z^{e_R}\right)^2\left(\left(g_Z^{f_L}\right)^2+\left(g_Z^{f_R}\right)^2\right)$$

(a) (b)

Fig. 6.9 Diagrams and the cross sections of $\left(e_L^- e^+ \to Z^0 \to f\bar{f}\right)$ and $\left(e_R^- e^+ \to Z^0 \to f\bar{f}\right)$. $Z^0 - f_L$ coupling and $Z^0 - f_R$ coupling are also different

6.5.1 Measurements of $\sin^2\theta_W$

An important parameter, $\sin^2\theta_W$ have been measured accurately using the following interactions:

$$e^- + e^+ \to Z^0 \to f + \bar{f}, \tag{6.54}$$

where the final state f shows a fermion. If f is a quark, the final state becomes hadronic events, $e^- + e^+ \to Z^0 \to q + \bar{q} \to$ hadron jets.

It is possible to polarize the initial electron spin to either right-handed helicity e_R^- or left-handed helicity e_L^- as shown in Fig. 6.9. The cross section of the reaction (6.54) for e_L^- or e_R^- are proportional to

$$\sigma_L \propto (g_Z^{e_L})^2\left(\left(g_Z^{f_L}\right)^2+\left(g_Z^{f_R}\right)^2\right), \quad \sigma_R \propto (g_Z^{e_R})^2\left(\left(g_Z^{f_L}\right)^2+\left(g_Z^{f_R}\right)^2\right), \tag{6.55}$$

respectively. Therefore, the asymmetry between the two cross sections is

$$A_{\text{LR}} = \frac{\sigma_L - \sigma_R}{\sigma_L + \sigma_R} = \frac{(g_Z^{e_L})^2 - (g_Z^{e_R})^2}{(g_Z^{e_L})^2 + (g_Z^{e_R})^2} = \frac{2(1 - 4\sin^2\theta_W)}{1 + (1 - 4\sin^2\theta_W)^2}. \tag{6.56}$$

This is called *Left-Right asymmetry*. The measurement of $\sin^2\theta_W$ from A_{LR} is a measurement of eight times of its deviation from 0.25. Since $\sin^2\theta_W(\sim 0.23)$

Table 6.2 Observable parameters and their measured values. α is the fine structure constant $= e^2/4\pi$. G_F is the Fermi constant measured from the muon lifetime

Obs. parameter	EW parameter	Measured Value	Unit
M_Z	$\frac{1}{2}v_0\sqrt{g^2+g'^2}$	91.1876 ± 0.0021	GeV/c^2
M_W	$\frac{1}{2}v_0 g$	80.385 ± 0.015	GeV/c^2
$\sin^2\theta_W$	$\frac{g'^2}{g^2+g'^2}$	0.23122 ± 0.00015	–
α	$\frac{g^2 g'^2}{4\pi(g^2+g'^2)}$	$1/(137.03599911 \pm 0.00000046)$	–
G_F	$\frac{1}{\sqrt{2}v_0^2}$	$(1.16637 \pm 0.00001)\times 10^{-5}$	$1/$GeV^2

happens to be close to 0.25, $\sin^2\theta_W$ can be measured accurately with this method. In addition, this measurement is just counting the number of events with the same detector and the systematic uncertainties are very small. The measurement of A_{LR} were performed by the SLD experiment in 1990s [1].

The Weinberg angle can also be measured from the Forward-Backward asymmetry A_{FB} of the final state fermions in the reactions Fig. 6.9,

$$A_{FB} = \frac{\sigma_f[\theta]+\sigma_f[\pi-\theta]}{\sigma_f[\theta]+\sigma_f[\pi-\theta]} = \left(\frac{2\cos\theta}{1+\cos^2\theta}\right)\left(\frac{2(1-4|Q_f|\sin^2\theta_W)}{1+(1-4|Q_f|\sin^2\theta_W)}\right),$$
(6.57)

where Q_f is the charge of the final state fermion f and θ is the angle between the Z^0 spin direction and the emission angle of f. Four experiments, ALEPH, DELPHI, OPAL and L3 at LEP accelerator measured $\sin^2\theta_W$ from A_{FB} by making use of the natural polarization of Z^0 ($P_Z = A_{LR} \sim 16\%$) caused by the cross-sectional difference of σ_L and σ_R [2]. Currently the Weinberg angle is measured as [3]

$$\sin^2\theta_W = 0.23121 \pm 0.00005 .$$
(6.58)

6.5.2 Test of the Electroweak Theory

There are three basic parameters g, g', and v_0 in the electroweak theory. On the other hand, five parameters that are combinations of the three basic parameters have been measured accurately as shown in Table 6.2. Since the number of measurements is more than the number of parameters, a strong tests of the theory can be performed.

G_F is the Fermi constant measured from the lifetime of the muon decay. The diagram of the muon decay is shown in Fig. 6.10 and the lifetime is parametrized by the Fermi constant,

$$G_F = \frac{1}{2\sqrt{2}}\frac{g_W^2}{M_W^2} = \frac{1}{\sqrt{2}v_0^2}.$$
(6.59)

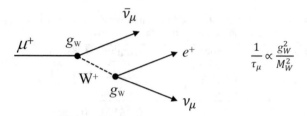

Fig. 6.10 μ^+ decay diagram. The lifetime $\tau_\mu \sim 2.2\,\mu s$ is inversely proportional to g_W^2/M_W^2, where the factor $1/M_W^2$ comes from the W^+ propagator

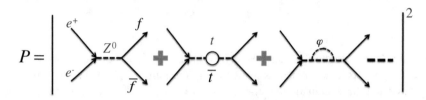

Fig. 6.11 Top quark and Higgs boson contributions to the radiative correction for $e^+ + e^- \rightarrow Z^0 \rightarrow f + \bar{f}$ reaction

Fig. 6.12 The relation of Higgs and the top quark masses and electroweak parameters. "Z-pole asymmetries" correspond to $\sin^2 \theta_W$. The bands width corresponds to 1σ. Figure 10.4 of [3]

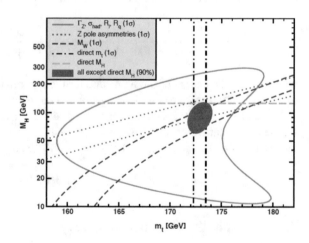

From the measured parameter values in Table 6.2, the three basic parameters are determined as

$$g \sim 0.65, \ g' \sim 0.36, \ v_0 \sim 246\text{GeV}. \tag{6.60}$$

The electroweak theory predicts that

$$\rho \equiv \frac{M_W^2}{M_Z^2 \cos^2 \theta_W} = \frac{4\pi\alpha}{\sqrt{2}G_F M_Z^2 \sin^2 2\theta_W} = 1. \tag{6.61}$$

If we put the measured value,

$$\frac{4\pi\alpha}{\sqrt{2}G_F M_Z^2 \sin^2 2\theta_W} = 0.9403 \pm 0.0007, \tag{6.62}$$

that is significantly smaller than unity. This discrepancy comes from radiative corrections of higher order diagrams shown in Fig. 6.11.

Historically, top quark mass and Higgs mass were predicted from this discrepancy and they were discovered at the predicted masses, which indicates the transitions and complicated mechanism of the electroweak theory described in this section is proven correct upto the higher order diagrams. Figure 6.12 shows the relation of Higgs and the top quark masses for various electroweak parameters. The expectation area from various parameters cross at a point and the the Higgs particle was discovered within the predicted region.

References

1. SLD collaboration: Phys. Rev. Lett. **84**, 5945 (2000)
2. ALEPH, DELPHI, L3, OPAL and SLD collaborations: Phys. Rep. **427**, 257 (2006)
3. Particle Data Group: Prog. Theor. Exp. Phys. **2020**, 083C01 (2020)

Fermion Mass and Chirality Oscillation

7

7.1 Introduction

In the standard model, right-handed fermion; f_R does not couple $W^\pm$. Therefore, f_R cannot decay by weak interactions. On the other hand, the right-handed muon; μ_R^- can be produced by high energy e^+-e^- electromagnetic annihilations,

$$e^+ + e^- \to \gamma^* \to \mu_R^- + \mu_L^+. \tag{7.1}$$

Experimentally, it has been observed that all the muons produced in the e^+e^- interaction decay weakly. Why the muon produced as μ_R^- can decay?

7.2 Chirality

The muon is a spin-1/2 fermion and its evolution in space-time is governed by the Dirac equation. The wave function of the muon is expressed generally by a four-component spinor,

$$\psi_\mu = \begin{pmatrix} u \\ v \end{pmatrix}, \tag{7.2}$$

where u and v themselves are two-component spinors,

$$u = \begin{pmatrix} u_1 \\ u_2 \end{pmatrix}, \quad v = \begin{pmatrix} v_1 \\ v_2 \end{pmatrix}. \tag{7.3}$$

The left-handed chirality (LHC) state to which the charged weak boson $W^\pm$ couples is defined as

© Springer Nature Switzerland AG 2021
F. Suekane, *Quantum Oscillations*, Lecture Notes in Physics 985,
https://doi.org/10.1007/978-3-030-70527-5_7

$$\psi_L \equiv \frac{1}{2}(1 - \gamma_5)\psi = \frac{1}{2}\begin{pmatrix} I & -I \\ -I & I \end{pmatrix}\begin{pmatrix} u \\ v \end{pmatrix} = \frac{1}{2}\begin{pmatrix} u - v \\ v - u \end{pmatrix} \tag{7.4}$$

and right-handed chirality (RHC) states to which $W^\pm$ does not couple is

$$\psi_R \equiv \frac{1}{2}(1 + \gamma_5)\psi = \frac{1}{2}\begin{pmatrix} I & I \\ I & I \end{pmatrix}\begin{pmatrix} u \\ v \end{pmatrix} = \frac{1}{2}\begin{pmatrix} u + v \\ u + v \end{pmatrix}, \tag{7.5}$$

where I is the 2×2 identity matrix and γ_5 is a product of all the Dirac matrices,

$$\gamma_5 = i\gamma_0\gamma_1\gamma_2\gamma_3 = \begin{pmatrix} 0 & I \\ I & 0 \end{pmatrix}. \tag{7.6}$$

We will use $\gamma_L \equiv (1 - \gamma_5)/2$, and $\gamma_R \equiv (1 + \gamma_5)/2$, hereafter.

In order to simplify the expression of equations, we define a mathematical expression such a way that if a spinor is multiplied to a matrix whose elements are scalar, the spinor is multiplied to each element of the matrix;

$$\psi_L = \frac{u - v}{2}\begin{pmatrix} 1 \\ -1 \end{pmatrix} \equiv c_L|L\rangle, \quad \psi_R = \frac{u + v}{2}\begin{pmatrix} 1 \\ 1 \end{pmatrix} \equiv c_R|R\rangle, \tag{7.7}$$

where

$$c_L = \frac{u - v}{\sqrt{2}}, \quad |L\rangle = \frac{1}{\sqrt{2}}\begin{pmatrix} 1 \\ -1 \end{pmatrix}, \quad c_R = \frac{u + v}{\sqrt{2}}, \quad |R\rangle = \frac{1}{\sqrt{2}}\begin{pmatrix} 1 \\ 1 \end{pmatrix}. \tag{7.8}$$

We regard $|R\rangle$ to be the basic vector of RHC state and $|L\rangle$, LHC. Any spin-1/2 wave function; ψ is a superposition of the LHC and RHC base states,

$$\psi[t] = c_R[t]|R\rangle + c_L[t]|L\rangle. \tag{7.9}$$

7.3   Dirac Equation as Chirality Transition Equation

Since the muon is a spin-1/2 fermion, the wave function of the muon satisfies the Dirac equation,

$$\gamma^\mu \partial_\mu \psi = -im\psi. \tag{7.10}$$

If we apply γ_L or γ_R from the left, the Dirac equation leads

$$\begin{cases} \gamma^\mu \partial_\mu \psi_R = -im\psi_L \\ \gamma^\mu \partial_\mu \psi_L = -im\psi_R, \end{cases} \tag{7.11}$$

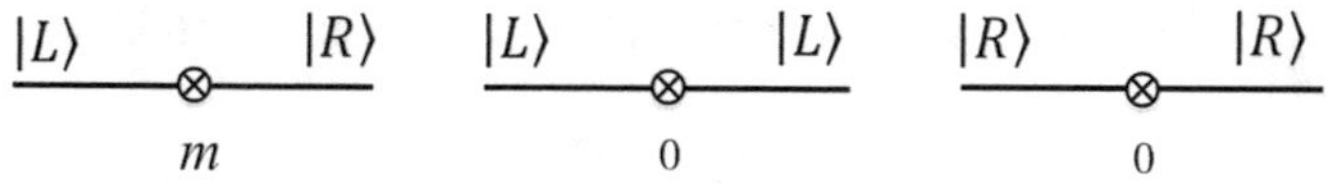

$$\begin{array}{ccccccc}
|L\rangle & & |R\rangle & |L\rangle & & |L\rangle & |R\rangle & & |R\rangle \\
\rule{3cm}{0.4pt}\!\!\otimes\!\!\rule{0pt}{0pt} & & & \rule{3cm}{0.4pt}\!\!\otimes\!\!\rule{0pt}{0pt} & & & \rule{3cm}{0.4pt}\!\!\otimes\!\!\rule{0pt}{0pt} \\
& m & & & 0 & & & 0
\end{array}$$

Fig. 7.1 Transition between the chirality state in the Dirac equation

where we used the relation $\gamma_L \gamma^\mu = \gamma^\mu \gamma_R$. The equation form (7.11) implies that the Dirac equation transforms $\psi_L \leftrightarrow \psi_R$. To see the transformation effect clearly, we think about the fermion at rest,

$$\frac{d}{dt}\psi = -im\gamma_0\psi .\tag{7.12}$$

If we explicitly write Eq. (7.12),

$$\begin{pmatrix} \dot{u} \\ \dot{v} \end{pmatrix} = -im \begin{pmatrix} u \\ -v \end{pmatrix}.\tag{7.13}$$

Using (7.8), the equation can be written as

$$\dot{c}_R|R\rangle + \dot{c}_L|L\rangle = -im(c_L|R\rangle + c_R|L\rangle) ,\tag{7.14}$$

where c_L and c_R are spinors. This leads the simultaneous differential equations of c_R and c_L,

$$\frac{d}{dt}\begin{pmatrix} c_R \\ c_L \end{pmatrix} = -im\begin{pmatrix} 0 & 1 \\ 1 & 0 \end{pmatrix}\begin{pmatrix} c_R \\ c_L \end{pmatrix}.\tag{7.15}$$

This means new $|L\rangle$-component is generated with amplitude $-im$ per unit time from $|R\rangle$, and vice versa. There are no self-transitions, $|L\rangle \leftrightarrow |L\rangle$, $|R\rangle \leftrightarrow |R\rangle$ and the transitions between $|L\rangle$ and $|R\rangle$ states can be shown as in Fig. 7.1.

Therefore, the Dirac equation is actually a chirality swapping equation. The general solution of this equations is

$$\begin{cases} c_R[t] = c_+e^{-imt} + c_-e^{imt} \\ c_L[t] = c_+e^{-imt} - c_-e^{imt} , \end{cases}\tag{7.16}$$

where $c_\pm = (c_R[0] \pm c_L[0])/2$ are integral constants to be determined by boundary conditions. The corresponding general wave function is

$$\psi[t] = c_+(|R\rangle + |L\rangle)e^{-imt} + c_-(|R\rangle - |L\rangle)e^{imt} .\tag{7.17}$$

From this, the mass eigenstates are

$$\psi_\pm[t] = \frac{1}{\sqrt{2}}(|R\rangle \pm |L\rangle)e^{\mp imt} .\tag{7.18}$$

If muon is produced as μ_R^- as in the interaction (7.1), the initial condition is

$$c_R[0] = u_0, \quad c_L[0] = 0 , \tag{7.19}$$

where u_0 is a unit spinor representing the initial spin direction of the μ_R^-. The wave function (7.17) becomes

$$\psi_R[t] = u_0(\cos[m_\mu t]|R\rangle - i \sin[m_\mu t]|L\rangle) . \tag{7.20}$$

Therefore, the probability that the fermion is in $|L\rangle$ state at time t is

$$P_{R\to L}[t] = |\langle L|\psi[t]\rangle|^2 = \sin^2[m_\mu t] . \tag{7.21}$$

This means that even if the muon is produced in the RHC state, LHC state is generated after $t \sim 1/m_\mu \sim 10^{-23}$ s after the production that is much shorter than the muon lifetime ($\sim 10^{-6}$ s). While the muon is in the LHC state, it can decay weakly.

7.4   Decay Effect

The effect of particle decay can be expressed by a wave function with imaginary mass component such as

$$\psi[t] = \psi[0] \exp[-i\tilde{m}t]; \quad \tilde{m} = m_0 - i\Gamma . \tag{7.22}$$

In this case the probability of the existence of the particle at time t is

$$P[t] = |\psi[t]|^2 = |\psi_0|^2 \exp[-2\Gamma t] . \tag{7.23}$$

This means that due to the decay, the probability of existence reduces to $e^{-2\Gamma t}$ after time t. Therefore, Γ corresponds to the mean life time as $\tau = 1/2\Gamma$.

The decay effect of the muon can be introduced by giving an imaginary amplitude to the μ_L self-transition as shown in Fig. 7.2, where Γ shows the strength of the muon decay. In the standard model, the muon mass is generated by the Yukawa coupling to the Higgs field. The Higgs field swaps the chirality as we saw in the previous section. The muon decay is caused by the coupling to the weak $W^\pm$ boson. Since $W^\pm$ boson couples only with the left-handed muon, there is no decay amplitude for μ_R.

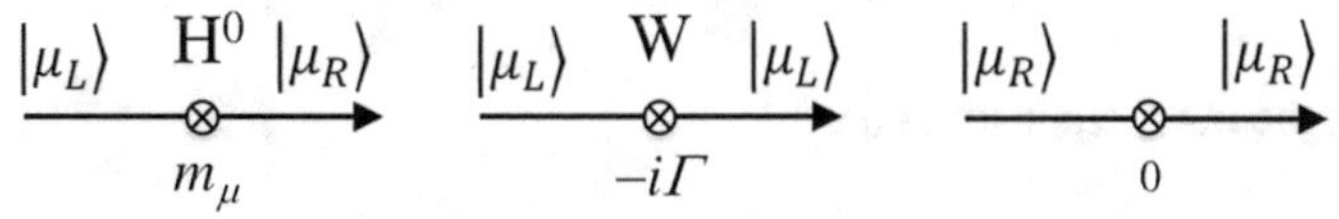

Fig. 7.2 The Higgs field swaps the chirality and generates the mass. The effect of weak decay of μ_L^- is introduced by putting an imaginary amplitude $-i\Gamma$ to the self-transition of μ_L^-

Γ is in the order of the inverse of the muon lifetime. Therefore, it is much smaller than the muon mass,

$$\Gamma \sim \frac{1}{2\tau_\mu} = \frac{1}{2 \times 2.2 \ \mu s} \sim 10^{-10} \ \text{eV} \ll m_\mu \sim 10^8 \ \text{eV} . \tag{7.24}$$

An arbitrary muon wave function can be written as

$$\Psi_\mu[t] = c_L[t]|\mu_L\rangle + c_R[t]|\mu_R\rangle . \tag{7.25}$$

From the transitions in Fig. 7.2, the transition equation is

$$\frac{d}{dt}\begin{pmatrix} c_L \\ c_R \end{pmatrix} = -i \begin{pmatrix} -i\Gamma & m_\mu \\ m_\mu & 0 \end{pmatrix}\begin{pmatrix} c_L \\ c_R \end{pmatrix} . \tag{7.26}$$

By solving the equations, the general wave function is

$$\Psi_\mu[t] = e^{-(\Gamma/2)t}\left((ae^{-im_\mu t} - be^{im_\mu t})|\mu_R\rangle + (ae^{-im_\mu t} + be^{im_\mu t})\left(|\mu_L\rangle - i\frac{\Gamma}{2m_\mu}|\mu_R\rangle\right)\right), \tag{7.27}$$

where $(\Gamma/m_\mu)^2$ terms are ignored. a and b are integral constants.

If the muon is right-handed at $t = 0$,

$$\Psi_\mu[0] = (a - b)|\mu_R\rangle + (a + b)\left(|\mu_L\rangle - i\frac{\Gamma}{2m_\mu}|\mu_R\rangle\right) = |\mu_R\rangle . \tag{7.28}$$

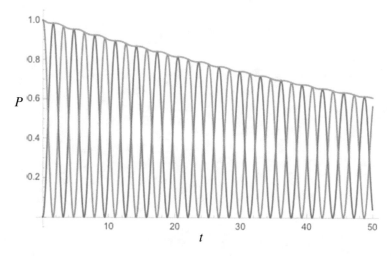

Fig. 7.3 Green line: probability to be μ^-. Orange and blue lines: μ_R^- and μ_L^- components, respectively. The μ_R^- and μ_L^- components oscillate in μ^-. The ripple on the $e^{-\Gamma t}$ envelop shows that the decay happens only while it is μ_L^- state. The oscillation frequency is exaggerated to slow to see the effect of the oscillation ($\Gamma/\omega = 0.01$ is assumed)

the integral constants are determined as, $(a = -b = u_0/2)$ and the wave function becomes

$$\Psi_\mu[t] = u_0 e^{-(\Gamma/2)t} \left(\left(\cos m_\mu t - \frac{\Gamma}{2m_\mu} \sin m_\mu t \right) |\mu_R\rangle - i \sin m_\mu t |\mu_L\rangle \right) . \quad (7.29)$$

The probability for the existence of the muon at time t is

$$P[t] = |\psi[t]|^2 = e^{-\Gamma t} \left(1 - \frac{\Gamma}{2m_\mu} \sin[2m_\mu t] \right) . \quad (7.30)$$

Since $\Gamma \ll 2m_\mu = \omega$,

$$P[t] \sim e^{-\Gamma t}. \quad (7.31)$$

Therefore, the exponential decay property is derived as expected.

The probability to be in the left-handed state is

$$|\langle \mu_L | \Psi[t] \rangle|^2 = e^{-\Gamma t} \sin^2 \omega t . \quad (7.32)$$

Figure 7.3 shows the muon decay and the effect of the $\mu_L^- $-$\mu_R^-$ oscillation.

Quark Mass, Cabibbo Angle and CKM Mixing Matrix

8

8.1 Introduction

In the standard model of the elementary particle physics, there are 3 charge $-(1/3)e$ quarks; (d, s, b) and 3 charge $+(2/3)e$ quarks; (u, c, t). We collectively call the (u, c, t) quarks as "Up-type quark" and the (d, s, b) quarks as "Down-type quark" in this chapter. We start the discussion using the four lighter quarks; (d, s) and (u, c) and introduce the Cabibbo angle θ_C. Then we will extend the discussion to six-quark system to introduce CKM (Cabibbo–Kobayashi–Maskawa) mixing later.

8.2 Four-Quark System and Cabibbo Angle, θ_C

In the standard model, the quarks change their flavors coupling with the charged weak bosons $(W^\pm)$ as shown in Fig. 8.1.

The corresponding weak interaction Hamiltonian is[1]

$$\mathcal{H}_{q'q'W} = g_W\{\overline{u'}\gamma^\mu d'\}W_\mu^+ + g_W\{\overline{c'}\gamma^\mu s'\}W_\mu^+ + \text{H.C.} , \qquad (8.1)$$

where the dash on q' is to distinguish the weak eigenstate from the mass eigenstate q. The definition of the weak eigenstate is that the transition between $(s' \leftrightarrow u')$ nor $(d' \leftrightarrow c')$ are not caused by $W^\pm$. The terms corresponding to the Hermitian Conjugate (H.C.) are reverse processes, such as $g_W\{\overline{d'}\gamma^\mu u'\}W_\mu^-$ and are ignored in the following discussions unless they become necessary.

---

[1] Each quark is in left-handed chirality state. See Chap. 7.

© Springer Nature Switzerland AG 2021
F. Suekane, *Quantum Oscillations*, Lecture Notes in Physics 985,
https://doi.org/10.1007/978-3-030-70527-5_8

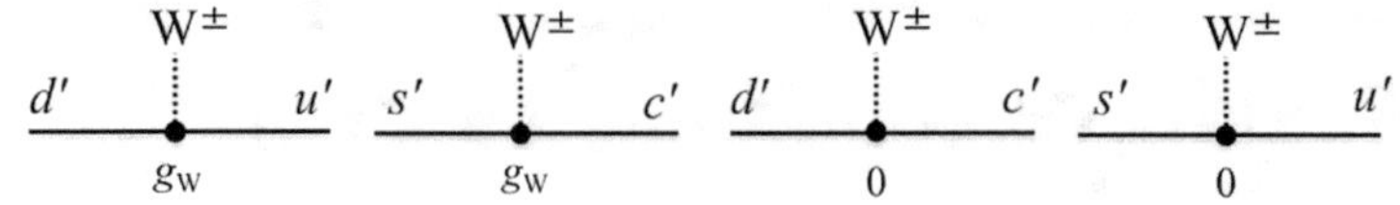

Fig. 8.1 Coupling between $W^{\pm}$ boson and weak eigenstate s', d', u', c' quarks. Dash (') indicates that these quarks are weak eigenstate

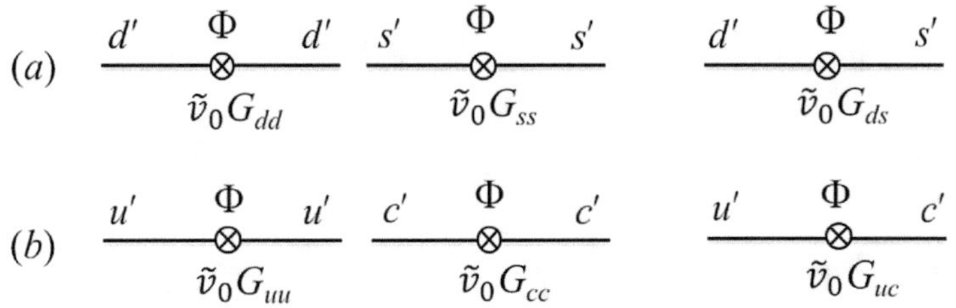

Fig. 8.2 Quark transitions due to the Higgs field, Φ. **a** Transitions between d' and s'. **b** Transitions between u' and c'. The transition amplitudes are proportional to the product of the vacuum expectation value ($\tilde{v}_0$) of the Higgs field and the corresponding coupling constant $G_{\alpha\beta}$

In the standard model, the quarks also couple to the Higgs field (Φ) and transform as shown in Fig. 8.2.[2] $\tilde{v}_0 = v_0/\sqrt{2} \sim 173\,\text{GeV}$ can be regarded as a potential energy which exists throughout our universe and transition amplitudes are proportional to $\tilde{v}_0$. Due to the cross-transition (G_{ds} term), the general wave function of d' and s' quark system is expressed as

$$\psi_D[t] = d'[t]|d'\rangle + s'[t]|s'\rangle . \tag{8.2}$$

From the transitions in Fig. 8.2a, the transition equation is

$$\frac{d}{dt}\begin{pmatrix} s' \\ d' \end{pmatrix} = -i\tilde{v}_0 \begin{pmatrix} G_{ss} & G_{ds} \\ G_{ds} & G_{dd} \end{pmatrix}\begin{pmatrix} s' \\ d' \end{pmatrix} . \tag{8.3}$$

We call the quark that has larger self-transition (G_{qq}) as s'-quark and smaller one as d'-quark. That means by definition, $G_{ss} > G_{dd}$. From Eqs. (B.35), the wave functions of the mass eigenstate are[3]

$$\begin{cases} |\psi_s\rangle = (\cos\theta_D|s'\rangle + \sin\theta_D|d'\rangle)e^{-im_s t} \equiv |s\rangle e^{-im_s t} \\ |\psi_d\rangle = (-\sin\theta_D|s'\rangle + \cos\theta_D|d'\rangle)e^{-im_d t} \equiv |d\rangle e^{-im_d t} . \end{cases} \tag{8.4}$$

The mixing angle θ_D is defined as in the mixing triangle Fig. 8.3a,

$$\tan 2\theta_D = \frac{2G_{ds}}{G_{ss} - G_{dd}} . \tag{8.5}$$

---

[2]See Chap. 6.
[3]We can safely assume $\phi = 0$ and $\delta_{\pm} = 0$ here.

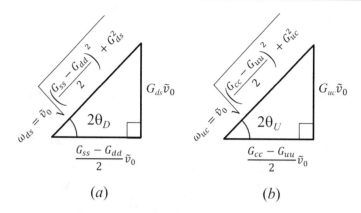

Fig. 8.3 Mixing triangles for **a** d' and s' quarks, and **b** u' and c' quarks in the Higgs potential

The quark masses are, from Eq. (B.36),

$$m_d = \overline{m}_D - \omega_{ds}, \quad m_s = \overline{m}_D + \omega_{ds}, \tag{8.6}$$

where

$$\overline{m}_D = \tilde{v}_0 \frac{G_{dd} + G_{ss}}{2}, \quad \omega_{ds} = \frac{\tilde{v}_0}{2}\sqrt{(G_{ss} - G_{dd})^2 + 4G_{ds}^2}. \tag{8.7}$$

The relation between the coefficients of mass and flavor eigenstates is

$$\begin{pmatrix} s \\ d \end{pmatrix} = \begin{pmatrix} \cos\theta_D & \sin\theta_D \\ -\sin\theta_D & \cos\theta_D \end{pmatrix} \begin{pmatrix} s' \\ d' \end{pmatrix}. \tag{8.8}$$

With exactly the same procedure, by writing the wave function of the Up-type quarks as

$$\psi_U[t] = u'[t]|u'\rangle + c'[t]|c'\rangle, \tag{8.9}$$

the transition equation for the Up-type quarks is written, from Fig. 8.2b, as

$$\frac{d}{dt}\begin{pmatrix} c' \\ u' \end{pmatrix} = -i\tilde{v}_0 \begin{pmatrix} G_{cc} & G_{uc} \\ G_{uc} & G_{uu} \end{pmatrix}\begin{pmatrix} c' \\ u' \end{pmatrix}. \tag{8.10}$$

The mass eigenstate of the Up-type quarks is, then,

$$\begin{cases} \psi_c = (\cos\theta_U|c'\rangle + \sin\theta_U|u'\rangle)e^{-im_c t} \equiv |c\rangle e^{-im_c t} \\ \psi_u = (-\sin\theta_U|c'\rangle + \cos\theta_U|u'\rangle)e^{-im_u t} \equiv |u\rangle e^{-im_u t}, \end{cases} \tag{8.11}$$

where the mixing angle and masses are

$$\tan 2\theta_U = \frac{2G_{uc}}{G_{cc} - G_{uu}}, \quad m_{c/u} = \overline{m}_U \pm \omega_{uc},$$

$$\overline{m}_U = \tilde{v}_0 \frac{G_{cc} + G_{uu}}{2}, \quad \omega_{uc} = \frac{\tilde{v}_0}{2}\sqrt{(G_{cc} - G_{uu})^2 + 4G_{uc}^2}. \tag{8.12}$$

The relation of the coefficients are

$$\begin{pmatrix} c \\ u \end{pmatrix} = \begin{pmatrix} \cos\theta_U & \sin\theta_U \\ -\sin\theta_U & \cos\theta_U \end{pmatrix} \begin{pmatrix} c' \\ u' \end{pmatrix} . \tag{8.13}$$

If we rewrite the $W^{\pm}$-quark interaction Hamiltonian (8.1) by using the energy eigenstate,

$$
\begin{aligned}
\mathcal{H}_{q'q'W} \to \mathcal{H}_{qqW} &= gw \begin{pmatrix} \{(\sin\theta_U\bar{c} + \cos\theta_U\bar{u})\gamma^{\mu}(\sin\theta_D s + \cos\theta_D d)\} \\ + \{(\cos\theta_U\bar{c} - \cos\theta_U\bar{u})\gamma^{\mu}(\cos\theta_D s - \sin\theta_D d)\} \end{pmatrix} W_{\mu}^{+} \\
&= gw \begin{pmatrix} \cos[\theta_D - \theta_U](\{\bar{u}\gamma^{\mu}d\} + \{\bar{c}\gamma^{\mu}s\}) \\ + \sin[\theta_D - \theta_U](\{\bar{u}\gamma^{\mu}s\} - \{\bar{c}\gamma^{\mu}d\}) \end{pmatrix} W_{\mu}^{+} .
\end{aligned}
\tag{8.14}
$$

This formula means that the mixing angle of the couplings to the W boson always appear as the difference; $\theta_D - \theta_U$, and it is impossible to measure the absolute value of θ_D and θ_U separately. We call this observable angle difference "the Cabibbo angle" and write it as θ_C,

$$\theta_C \equiv \theta_D - \theta_U . \tag{8.15}$$

Since the absolute value of the mixing angles are not observable, we can define them. Usually we define $\theta_U = 0$ and consider $\theta_C = \theta_D$. This definition is equivalent to setting $G_{cu} = 0$ and Eq. (8.11) becomes

$$|c'\rangle = |c\rangle, \quad |u'\rangle = |u\rangle . \tag{8.16}$$

Therefore, usually, we consider as if the mass eigenstate and weak eigenstate of the Up-type quarks are the same and all the mixing effects are attributed to the Down-type quarks. This is pictorially shown in Fig. 8.4 using the mixing triangles. From now on, we will use this basis. Using the Cabibbo angle, Eq. (8.8) can be written as follows,

$$\begin{pmatrix} s' \\ d' \end{pmatrix} = \begin{pmatrix} \cos\theta_C & -\sin\theta_C \\ \sin\theta_C & \cos\theta_C \end{pmatrix} \begin{pmatrix} s \\ d \end{pmatrix} . \tag{8.17}$$

The Hamiltonian (8.14) indicates that at the mass eigenstates, $(s \leftrightarrow u)$ and $(d \leftrightarrow c)$ transitions are possible through the charged current weak interaction as shown in Fig. 8.5.

When we talk about quarks in hadrons, usually they are mass eigenstate because we distinguish hadrons based on their masses. Therefore, s-quark in a hadron can transform to u-quark emitting the W^- boson and forms another hadron. This is the reason that the K^- meson composed of $(s\bar{u})$ quarks can decay weakly although there is no coupling between s' and u' quarks as shown in Fig. 8.6.

$$K^- \to \pi^0 + W^- \tag{8.18}$$

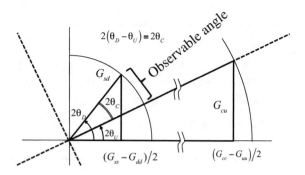

Fig. 8.4 The relations among mixing triangles for Down-type quarks, Up-type quarks, and the Cabibbo angle θ_C

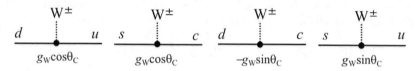

Fig. 8.5 Couplings between $W^\pm$ and mass eigenstate quarks; (u, c) and (d, s)

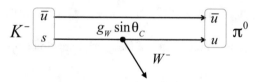

Fig. 8.6 In the decay of K^-, s-quark emits W^- and becomes u-quark, forming π^0 meson in the final state

The probability of the decay is proportional to $\sin^2 \theta_C$,

$$\Gamma_{K \to \pi} \propto g_W^2 \sin^2 \theta_C , \tag{8.19}$$

and $\tan^2 \theta_C$ times weaker than $d \to u$ decays if other conditions are the same and it is possible to measure the Cabibbo angle experimentally. Currently, the Cabibbo angle is measured as

$$\theta_C \sim 0.227 . \tag{8.20}$$

The original transition amplitudes are expressed by the observable parameters as

$$G_{ds} = \frac{m_s - m_d}{2\tilde{v}_0} \sin 2\theta_C, \qquad \begin{cases} G_{ss} = (\cos^2 \theta_C m_s + \sin^2 \theta_C m_d)/\tilde{v}_0 \\ G_{dd} = (\sin^2 \theta_C m_s + \cos^2 \theta_C m_d)/\tilde{v}_0 , \end{cases} \tag{8.21}$$

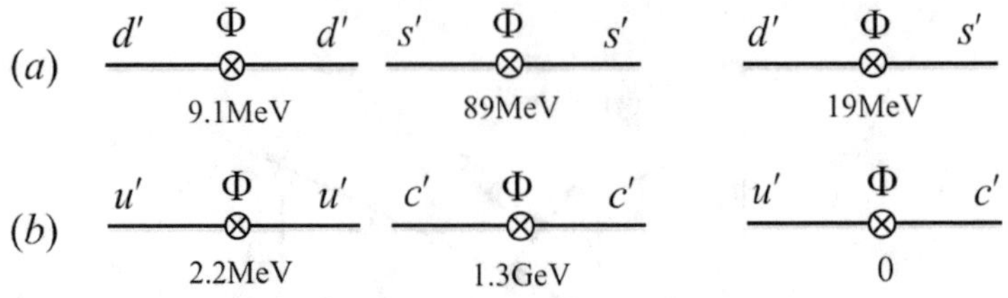

Fig. 8.7 Measured quark transition amplitudes; $G_{\alpha\beta}\tilde{v}_0$

The measured quark masses and the Higgs vacuum expectation value are

$$m_d \sim 4.7\,\mathrm{MeV}, \; m_s \sim 93\,\mathrm{MeV}, \; m_u \sim 2.2\,\mathrm{MeV}, \; m_c \sim 1.27\,\mathrm{GeV}, \; \tilde{v}_0 \sim 173\,\mathrm{GeV}. \tag{8.22}$$

From these observable information, the coupling constants of the Down-type quarks to the Higgs potential can be determined as

$$G_{ds} = 1.1 \times 10^{-4}, \quad G_{dd} = 5.3 \times 10^{-5}, \quad G_{ss} = 3.6 \times 10^{-4} ==> 5.1 \times 10^{-4}. \tag{8.23}$$

For Up-type quarks, we defined $G_{uc} = 0$ and

$$G_{uu} = m_u/\tilde{v}_0 \sim 1.3 \times 10^{-5}, \quad G_{cc} = m_c/\tilde{v}_0 \sim 7.3 \times 10^{-3}. \tag{8.24}$$

Figure 8.7 shows the measured values of the quark transition amplitudes.

8.2.1   Quark Flavor Oscillation

By using the Cabibbo angle $\theta_C = \theta_D - \theta_U$, the interaction Hamiltonian (8.14) can be rewritten as

$$\begin{aligned} \mathcal{H}_{qq'W} &= g_W W_\mu^- \left\{ \bar{u}\gamma^\mu (\cos\theta_C d + \sin\theta_C s) \right\} + \left\{ \bar{c}\gamma^\mu (\cos\theta_C s - \sin\theta_C d) \right\} \\ &= g_W W_\mu^- \left(\left\{ \bar{u}\gamma^\mu d' \right\} + \left\{ \bar{c}\gamma^\mu s' \right\} \right). \end{aligned} \tag{8.25}$$

This means, there are no $(c \leftrightarrow d')$ nor $(u \leftrightarrow s')$ transitions as shown in Fig. 8.8. From these relations, it is possible to produce a pure d' state in the decay of τ^- lepton as follows:

$$\tau^- \to \nu_\tau + W^- \mapsto \bar{u} + d'. \tag{8.26}$$

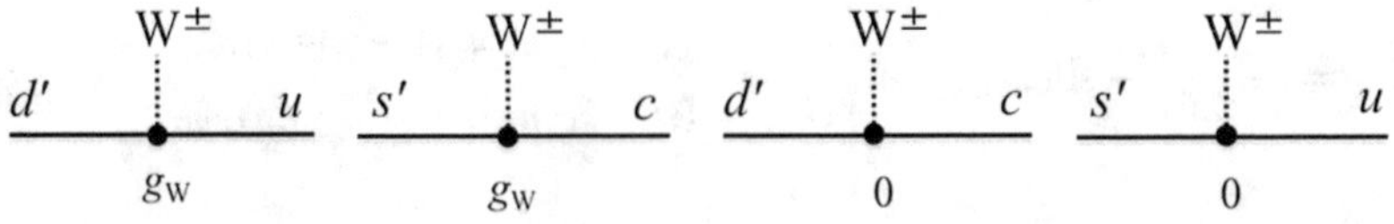

Fig. 8.8 Transitions between the Up-type quarks and the weak eigenstate of the Down-type quarks. There are no $(c \leftrightarrow d')$ nor $(u \leftrightarrow s')$ transitions

The amplitude of $\tau^- \to \nu_\tau + W^- \mapsto \bar{u} + s'$ decay is 0 and $\tau^- \to \nu_\tau + \bar{c} + s'$ is forbidden because of the large c-quark mass ($m_\tau < m_D(\bar{c}q)$).

If pure d' is produced at time $t = 0$, the initial condition for the wave function (8.2) is

$$d'[0] = 1, \quad s'[0] = 0 \tag{8.27}$$

and the quark state after time t is, from Eq. (B.26),

$$
\begin{pmatrix} s'[t] \\ d'[t] \end{pmatrix} = e^{-i\overline{m}_D t} \left(\cos[\omega_{ds} t] I - i \sin[\omega_{ds} t] \begin{pmatrix} \cos 2\theta_C & \sin 2\theta_C \\ \sin 2\theta_C & -\cos 2\theta_C \end{pmatrix} \right) \begin{pmatrix} 0 \\ 1 \end{pmatrix}
$$
$$
= e^{-i\overline{m}_D t} \begin{pmatrix} -i \sin[\omega_{ds} t] \sin 2\theta_C \\ \cos[\omega_{ds} t] + i \sin[\omega_{ds} t] \sin 2\theta_C \end{pmatrix}. \tag{8.28}
$$

The wave function at time t is

$$\psi_q[t] = e^{-i\overline{m}_D t} ((\cos[\omega_{ds} t] + i \sin[\omega_{ds} t] \cos 2\theta_C)|d'\rangle - i \sin[\omega_{ds} t] \sin 2\theta_C |s'\rangle). \tag{8.29}$$

Therefore, the probability that the s' quark is generated at time t is

$$P_{d' \to s'}[t] = |\langle s' | \psi_q[t] \rangle|^2 = \sin^2 2\theta_C \sin^2[\omega_{ds} t]. \tag{8.30}$$

Therefore, the $d' \leftrightarrow s'$ oscillation is taking place and the probability to be d' state or s' state is a function of time.

The wave function (8.29) can also be expressed by the mass eigenstate $|d\rangle$ and $|s\rangle$,

$$|\psi_q[t]\rangle = \cos \theta_C |d\rangle e^{-im_d t} + \sin \theta_C |s\rangle e^{-im_s t}. \tag{8.31}$$

This means the decay probability to $\tau^- \to \pi^-(\bar{u}d) + \nu_\tau$ is proportional to $\cos^2 \theta_C$ and that to $\tau^- \to K^-(\bar{u}s) + \nu_\tau$ is proportional to $\sin^2 \theta_C$.

8.2.2 Uncertainty Principle

Now we assume that we perform an experiment and we identify that τ^- decayed to π^- at time $t = 0$. In this case, the wave function of the Down-type quark part is

$$\psi_q[0] = |d\rangle = \cos \theta_C |d'\rangle - \sin \theta_C |s'\rangle. \tag{8.32}$$

After time t, the wave function becomes

$$\psi_q[t] = |d\rangle e^{-im_d t} = (\cos \theta_C |d'\rangle - \sin \theta_C |s'\rangle) e^{-im_d t}. \tag{8.33}$$

Therefore, the probability to be d' state or s' state does not depend on the time and the $d' \leftrightarrow s'$ oscillation described in the previous subsection seems to freeze. But why the oscillation freezes? Actually, the oscillations are still taking place for both b' and s' quarks,

$$|d'\rangle \xrightarrow{t} e^{-i\overline{m}_D t}((\cos[\omega_{ds}t] + i\sin[\omega_{ds}t]\cos 2\theta_C)|d'\rangle - i\sin[\omega_{ds}t]\sin 2\theta_C|s'\rangle)$$

$$|s'\rangle \xrightarrow{t} e^{-i\overline{m}_D t}((\cos[\omega_{ds}t] - i\sin[\omega_{ds}t]\cos 2\theta_C)|s'\rangle - i\sin[\omega_{ds}t]\sin 2\theta_C|d'\rangle) .$$

$$\tag{8.34}$$

However, for the particular state, $|d\rangle = \cos\theta_C|d'\rangle - \sin\theta_C|s'\rangle$, the $d' \to s'$ oscillation and $s' \to d'$ oscillation amplitudes cancels and it seems there is no oscillation,

$$\cos\theta_C|d'\rangle - \sin\theta_C|s'\rangle \xrightarrow{t} (\cos\theta_C|d'\rangle - \sin\theta_C|s'\rangle)e^{-i\overline{m}_D t} . \tag{8.35}$$

If our apparatus of the experiment can distinguish s quark and d quark by their mass difference, the energy resolution (δE) is smaller than the mass difference,

$$\delta E < m_s - m_d \sim 10^2 \,\text{MeV} . \tag{8.36}$$

This is technically quite reasonable assumption. Due to the uncertainty principle, the time resolution (δt) of this system is worse than

$$\delta t > \frac{1}{\delta E} > \frac{1}{m_s - m_d} = \frac{1}{2\omega_{ds}} \sim 10^{-23} \,\text{s} . \tag{8.37}$$

This means that the time resolution is worse than the oscillation period and it is impossible in principle to observe the oscillation pattern. We can only observe the averaged values which correspond to the physical effect of (8.33).

8.3   Six-Quark System

For six flavors (three Down-type quarks (d, s, b) and three Up-type quarks (u, c, t)), the transition effects become much more complicated. Again, we define that the Up-type quarks are mass eigenstate. The transitions of Up-type quarks are shown in Fig. 8.9.

The transitions between Down-type quarks are shown in Fig. 8.10. The Down-type quarks are mixed by the cross-transitions and the general wave function is expressed as

$$\psi_D[t] = d'[t]|d'\rangle + s'[t]|s'\rangle + d'[t]|b'\rangle . \tag{8.38}$$

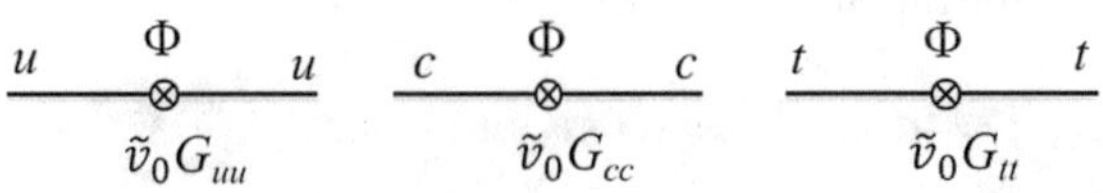

Fig. 8.9 Transitions among u, c and t quarks. They are defined to be mass eigenstate and there is no cross-transitions

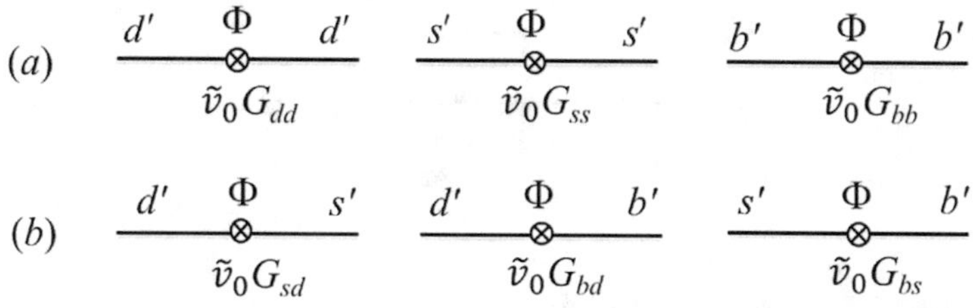

Fig. 8.10 Transitions among d', s' and b' quarks. **a** Self-transition. **b** Cross-transition

The transition equation which controls the time development of the coefficients is[4]

$$\frac{d}{dt}\begin{pmatrix} d' \\ s' \\ b' \end{pmatrix} = -i\tilde{v}_0 \begin{pmatrix} G_{dd} & G_{sd}^* & G_{bd}^* \\ G_{sd} & G_{ss} & G_{bs}^* \\ G_{bd} & G_{bs} & G_{bb} \end{pmatrix} \begin{pmatrix} d' \\ s' \\ b' \end{pmatrix}. \tag{8.39}$$

As a result of the transition equation, three mass eigenstates are obtained,

$$\begin{cases} \psi_d[t] = |d\rangle e^{-im_d t} \\ \psi_s[t] = |s\rangle e^{-im_s t} \\ \psi_b[t] = |b\rangle e^{-im_b t} . \end{cases} \tag{8.40}$$

From the analogy with the two flavor case (8.17), the mixing matrix between the mass eigenstates and the weak eigenstates can generally be expressed as

$$\psi' = \begin{pmatrix} d' \\ s' \\ b' \end{pmatrix} = \begin{pmatrix} V_{ud} & V_{us} & V_{ub} \\ V_{cd} & V_{cs} & V_{cb} \\ V_{td} & V_{ts} & V_{tb} \end{pmatrix} \begin{pmatrix} d \\ s \\ b \end{pmatrix} = V_{CKM}\psi . \tag{8.41}$$

This mixing matrix is called Cabibbo–Kobayashi–Maskawa (CKM) matrix. The CKM matrix elements are conventionally expressed as, for example, V_{us} instead of V_{ds}. This is because this element was measured from $s \rightarrow u$ decay probability.

$V_{\alpha\beta}$ and the quark masses m_β are functions of $G_{\alpha\beta}$. However, unlike the two flavor case in which the relations between transition amplitudes and mixing parameters & masses are simple as shown in Eqs. (8.5)~(8.7), for the three flavor case, $V_{\alpha\beta}$ and m_β are very complicate functions of $G_{\alpha\beta}$.[5]

Due to the conservation of the probability, using the expression Eq. (8.41),

$$|\psi'|^2 = \psi^\dagger \left(V_{CKM}^\dagger V_{CKM} \right) \psi = |\psi|^2 \tag{8.42}$$

---

[4]Traditionally, the lightest quark (d) is written at the top of the matrix and the heaviest quark (b) is written at the bottom.
[5]They consist of hundreds of terms of complicated combinations of $G_{\alpha\beta}$.

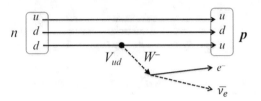

Fig. 8.11 Feynman diagram of the β decay

is required. Therefore, V_{CKM} is a unitary matrix and can be expressed by three real and one imaginary parameters [2]. One way to parametrize the mixing matrix is, using $\theta_{12}, \theta_{13}, \theta_{23}$ and $e^{i\delta}$,

$$V_{CKM} = \begin{pmatrix} 1 & 0 & 0 \\ 0 & c_{23} & s_{23} \\ 0 & -s_{23} & c_{23} \end{pmatrix} \begin{pmatrix} c_{13} & 0 & s_{13}e^{-i\delta} \\ 0 & 1 & 0 \\ -s_{13}e^{i\delta} & 0 & c_{13} \end{pmatrix} \begin{pmatrix} c_{12} & s_{12} & 0 \\ -s_{12} & c_{12} & 0 \\ 0 & 0 & 1 \end{pmatrix}$$

$$\tag{8.43}$$

$$= \begin{pmatrix} c_{12}c_{13} & s_{12}c_{13} & s_{13}e^{-i\delta} \\ -s_{12}c_{23} - c_{12}s_{23}s_{13}e^{i\delta} & c_{12}c_{23} - s_{12}s_{23}s_{13}e^{i\delta} & s_{23}c_{13} \\ s_{12}s_{23} - c_{12}c_{23}s_{13}e^{i\delta} & -s_{23}c_{12} - s_{12}c_{23}s_{13}e^{i\delta} & c_{23}c_{13} \end{pmatrix},$$

where $c_{ij} = \cos\theta_{ij}$ and $s_{ij} = \sin\theta_{ij}$.

8.3.1 Measurement of the CKM Matrix Elements

The CKM matrix elements have been measured as follows.[6]

8.3.1.1 V_{ud}
V_{ud} is measured from nuclear β decays. The elementary process of the β decay is

$$d \rightarrow u + W^- \rightarrow u + e^- + \bar{\nu}_e \tag{8.44}$$

and the diagram is shown in Fig. 8.11. This decay width is proportional to $|V_{ud}|^2$ and the mixing matrix element is measured as

$$|V_{ud}| \sim 0.97370 \pm 0.00014 . \tag{8.45}$$

---

[6]The values and explanations are taken from [1].

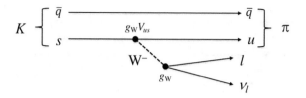

Fig. 8.12 Feynman diagram of $K \to \pi + l^- + \nu_l$ decay

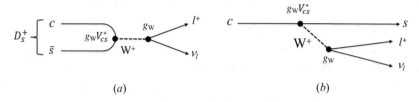

(a)   *(b)*

Fig. 8.13 a Feynman diagram of $D_s^+ \to l^+ + \nu_l$. **b** elementary process of $c \to s + l^+ + \nu_l$ decay

8.3.1.2 V_{us}

V_{us} is known as the Cabibbo angle in the two flavor model. This element is measured mainly by the decay widths of the $K \to \pi + l + \nu$ decays. The diagram of the decay is shown in Fig. 8.12. The elementary process is

$$s \to u + W^- \to u + l^- + \nu_l . \tag{8.46}$$

The measured value is

$$|V_{us}| = 0.2245 \pm 0.0008 . \tag{8.47}$$

8.3.1.3 V_{cs}

V_{cs} is measured from D_s decay lifetime and $D \to K + l + \nu$ decay width. The diagrams are shown in Fig. 8.13. The elementary process for those interactions are

$$\begin{aligned} c + \bar{s} &\to W^+ \to l^+ + \nu_l , \\ c &\to s + W^- \to s + l^+ + \nu_l , \end{aligned} \tag{8.48}$$

respectively. The measured value is

$$|V_{cs}| = 0.987 \pm 0.011 . \tag{8.49}$$

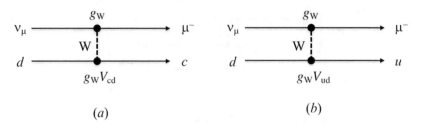

Fig. 8.14 Feynman diagram of **a** $\nu_\mu + d \rightarrow \mu^- + c$ and **b** $\nu_\mu + d \rightarrow \mu^- + u$

8.3.1.4 V_{cd}

V_{cd} is measured by the cross section of $\nu_\mu + d \rightarrow \mu^- + c$ and $\nu_\mu + d \rightarrow \mu^- + u$ scatterings. The Feynman diagrams of the scatterings are shown in Fig. 8.14. The ratio is

$$\frac{\sigma(\nu_\mu + d \rightarrow \mu^- c)}{\sigma(\nu_\mu + d \rightarrow \mu^- u)} \propto \frac{|V_{cd}|^2}{|V_{ud}|^2} . \tag{8.50}$$

From this, V_{cd} is measured as

$$|V_{cd}| = 0.221 \pm 0.004 . \tag{8.51}$$

8.3.1.5 V_{cb}

V_{cb} is measured by the decay width of $B \rightarrow D + X$. Figure 8.15 shows the diagram of an example decay; $B^- \rightarrow D^0 + l^- + \bar{\nu}$. The elementary process is

$$b \rightarrow c + W^- \tag{8.52}$$

and the measured value is

$$|V_{cb}| = (4.10 \pm 0.14) \times 10^{-2} . \tag{8.53}$$

8.3.1.6 V_{ub}

V_{ub} is measured by the decay width of $B \rightarrow h(u) + l + \nu$, where $h(u)$ means hadrons which contain the leading u-quark. Figure 8.16 shows an example of this decay diagram.

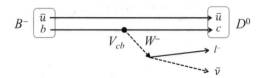

Fig. 8.15 Feynman diagram of $B^- \rightarrow D^0 + l^- + \bar{\nu}$

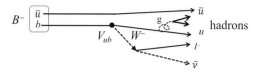

Fig. 8.16 Feynman diagram of $B^- \to h(u)$

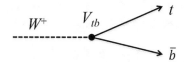

Fig. 8.17 Diagram of $W^+ \to t + \bar{b}$ decay

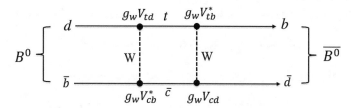

Fig. 8.18 Feynman diagram of B^0-$\overline{B^0}$ oscillation. The amplitude of this diagram is proportional to $V_{tb}V_{tb}^*V_{cb}^*V_{cd}$

The measured value is

$$|V_{cb}| = (3.82 \pm 0.24) \times 10^{-3} . \tag{8.54}$$

8.3.1.7 V_{tb}

V_{tb} is measured by $pp(\text{or } p\bar{p}) \to W^* + X \to t + \bar{b} + X$. Figure 8.17 shows an example of this decay diagram. The elementary process is

$$W^+ \to t + \bar{b} \tag{8.55}$$

and the measured value is,

$$|V_{tb}| = 1.013 \pm 0.030 . \tag{8.56}$$

8.3.1.8 V_{ts}, V_{td}

V_{ts} and V_{td} are measured from the process of the B^0-$\overline{B^0}$ oscillations. For example, the amplitude of the Fig. 8.18 is proportional to $V_{tb}V_{tb}^*V_{cb}^*V_{cd}$, and the measured values are

$$|V_{td}| = (8.0 \pm 0.3) \times 10^{-3}, \quad |V_{ts}| = (3.88 \pm 0.11) \times 10^{-2} . \tag{8.57}$$

The ratio of them is measured more precisely as

$$\frac{|V_{td}|}{|V_{ts}|} = 0.205 \pm 0.006 . \tag{8.58}$$

8.3.1.9 $e^{i\delta}$

δ is related to the CP violation and it is measured through CP violation of K^0 and B^0 system. The detail is described in Chap. 9. The δ has been measured as follows.

$$\delta \sim 1.20 \pm 0.05 . \tag{8.59}$$

8.3.1.10   Matrix Elements of V_{CKM}

Putting together the measured values, the mixing angles are determined as

$$\theta_{12} \sim 0.229, \quad \theta_{23} \sim 0.0405, \quad \theta_{13} \sim 0.0036 . \tag{8.60}$$

Using (8.59) and (8.60), the CKM matrix (8.43) is calculated as

$$V_{\text{CKM}} \sim \begin{pmatrix} 0.974 & 0.227 & 0.0013 - 0.0034i \\ -0.227 & 0.973 & 0.0405 \\ 0.00792 - 0.0033i & -0.0397 & 1.00 \end{pmatrix} . \tag{8.61}$$

8.3.2   Transition Amplitude $G_{\alpha\beta}$

If we recall how the quark masses and mixings are derived from the transition amplitude, the mathematical process is as follows. The transition equation (8.39) is written as

$$\frac{d}{dt}\psi' = -iG\psi' , \tag{8.62}$$

where G is the transition amplitude matrix. Then we assume there is some unitary matrix U which diagonalize G,

$$\frac{d}{dt}(U\psi') = -i(UGU^\dagger)(U\psi') . \tag{8.63}$$

$M \equiv UGU^\dagger$ is called the quark mass matrix. Correspondingly, $\psi = U\psi'$ is the mass eigenstate of the quarks. Comparing with the definition of V_{CKM}, Eq. (8.41),

$$V_{\text{CKM}} = U^{-1} = U^\dagger . \tag{8.64}$$

The transition amplitudes can be obtained from

$$G = U^\dagger MU = V_{\text{CKM}} M V_{\text{CKM}}^\dagger . \tag{8.65}$$

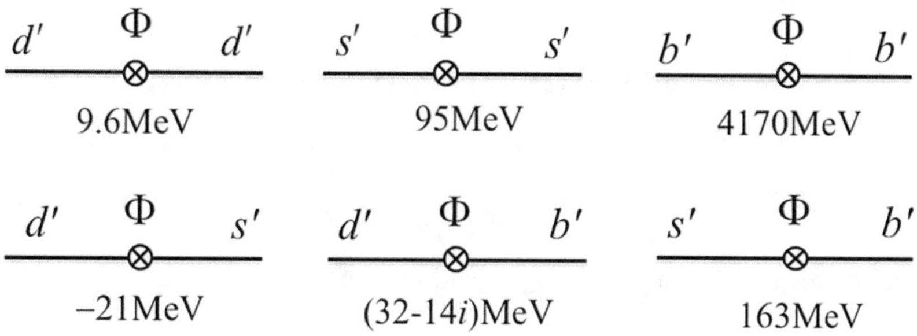

Fig. 8.19 Transition amplitudes of D-type quarks

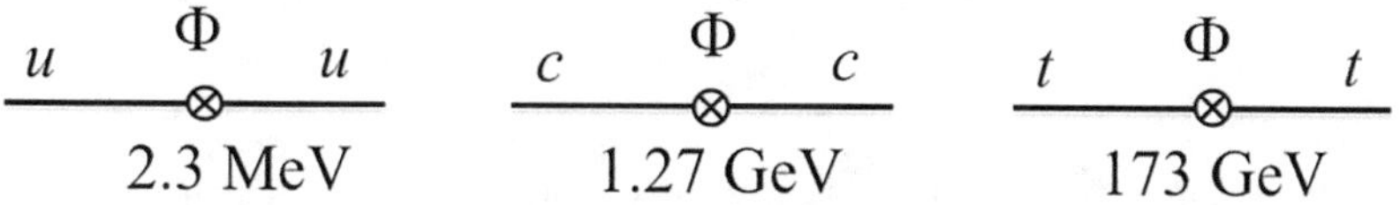

Fig. 8.20 Transition amplitudes of U-type quarks. The cross-transition amplitude is defined to be 0

$\mathcal{M}$ is, from quark mass measurements,

$$\mathcal{M} = \begin{pmatrix} m_d & 0 & 0 \\ 0 & m_s & 0 \\ 0 & 0 & m_b \end{pmatrix} \sim \begin{pmatrix} 4.7 & 0 & 0 \\ 0 & 93 & 0 \\ 0 & 0 & 4180 \end{pmatrix} \text{MeV} . \tag{8.66}$$

$\mathcal{G}$ is determined using (8.61) and (8.66) as

$$\mathcal{G} \sim \begin{pmatrix} 9.6 & -21 & 32 + 14i \\ -21 & 95 & -162 \\ 32 - 14i & -163 & 4170 \end{pmatrix} \text{MeV} . \tag{8.67}$$

Figure 8.19 shows the D-type quark transition amplitudes.

For U-type quarks, the cross-transition amplitudes are defined as 0 and the self-transition amplitudes are shows in Fig. 8.20.

The top quark-Higgs coupling constant,

$$G_{tt} = \frac{173 \,[\text{GeV}]}{\tilde{v}_0} \sim 0.99 \tag{8.68}$$

is surprisingly close to the unity.

8.3.3 Quark Flavor Oscillation

The three flavor oscillation probabilities are [2]

$$P[\alpha' \to \beta'] = \delta_{\alpha\beta} - 4 \sum_{k>j} \Re\left[\Omega_{kj}^{\alpha\beta}\right] \sin^2\left[\frac{\Delta m_{kj} t}{2}\right] - 2 \sum_{k>j} \Im\left[\Omega_{kj}^{\alpha\beta}\right] \sin\left[\Delta m_{kj} t\right] , \tag{8.69}$$

where

$$\Omega_{kj}^{\alpha\beta} = V_{\alpha k} V_{\beta k}^* V_{\alpha j}^* V_{\beta j}, \quad \Delta m_{ij} = m_i - m_j . \tag{8.70}$$

Due to the CPT invariance,

$$P[\overline{\beta'} \to \overline{\alpha'}] = P[\alpha' \to \beta']. \tag{8.71}$$

Therefore, the oscillation probability of anti-quarks are

$$
\begin{aligned}
P[\overline{\alpha'} \to \overline{\beta'}] &= P[\beta' \to \alpha'] \\
&= \delta_{\alpha\beta} - 4 \sum_{k>j} \Re \left[\Omega_{kj}^{\alpha\beta} \right] \sin^2 \left[\frac{\Delta m_{kj} t}{2} \right] + 2 \sum_{k>j} \Im \left[\Omega_{kj}^{\alpha\beta} \right] \sin \left[\Delta m_{kj} t \right]
\end{aligned}
\tag{8.72}
$$

An essential difference from the two flavor oscillations is that the imaginary component of the transition amplitude is observable in the three flavor oscillations. The difference of the oscillation probability between the quarks (8.69) and anti-quarks (8.72),

$$\Delta P = 4 \sum_{k>j} \Im \left[\Omega_{kj}^{\alpha\beta} \right] \sin \left[\Delta m_{kj} t \right] \tag{8.73}$$

is generated from the imaginary component of the mixing matrix element which come from the imaginary component of the transition amplitudes.

References

1. Particle Data Group, Prog. Theor. Exp. Phys. **2020**, 083C01 (2020)
2. Suekane, F.: Neutrino Oscillations; A Practical Guide to Basics and Applications. Springer, Berlin (2015)

Part III
Weak Interactions

K^0-$\overline{K^0}$ Oscillation and CP Violation

9.1   Introduction

Probably the K^0-$\overline{K^0}$ system is the most famous example of quantum oscillations. However, the process actually contains very profound properties, and we can learn many important physics from it. For example, the CP violation was first observed in this system and explained by K^0-$\overline{K^0}$ oscillation. In the K^0-$\overline{K^0}$ system, oscillation between CP eigenstates takes place, that means the CP quantum number changes in time. This is observed as a CP-violating effect. In this section, the physics of the K^0-$\overline{K^0}$ system is described from the oscillation and mixing point of view.

9.2 K^0-$\overline{K^0}$ Oscillation and Prediction of the Charm Quark Mass

The quark structures of K^0 and $\overline{K^0}$ are

$$\left|K^0\right\rangle = |d\bar{s}\rangle , \quad \left|\overline{K^0}\right\rangle = |\bar{d}s\rangle . \tag{9.1}$$

At first glance, the K^0 and $\overline{K^0}$ seem to be a mass eigenstate, like π^0, ρ^0, etc. However, there is a mutual transition between K^0 and $\overline{K^0}$ by diagrams shown in Fig. 9.1. The coupling structure comes from the relations shown in Fig. 8.5. There are also diagrams in which c or t quark intermediates instead of u quark. However, we start with only the intermediate u quark as shown in Fig. 9.1 to simplify this introductory discussion.

Due to this $K^0 \leftrightarrow \overline{K^0}$ transition, the general wave function of the neutral K system is expressed as a superposition of K^0 and $\overline{K^0}$ as

$$|\psi_K[t]\rangle = C_K[t]\left|K^0\right\rangle + C_{\overline{K}}[t]\left|\overline{K^0}\right\rangle . \tag{9.2}$$

© Springer Nature Switzerland AG 2021
F. Suekane, *Quantum Oscillations*, Lecture Notes in Physics 985,
https://doi.org/10.1007/978-3-030-70527-5_9

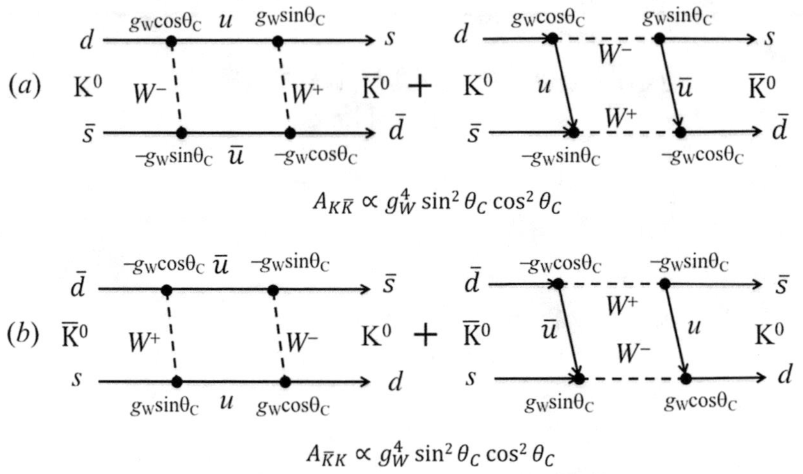

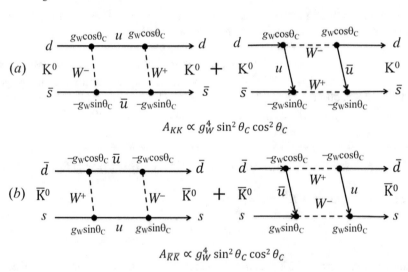

Fig. 9.1 $K^0 \leftrightarrow \overline{K^0}$ transition through intermediate u quarks. **a** $K^0 \rightarrow \overline{K^0}$. **b** $\overline{K^0} \rightarrow K^0$. θ_C is the Cabibbo angle

Fig. 9.2 Self-transitions of K^0 and $\overline{K^0}$ through u quarks. **a** $K^0 \rightarrow K^0$. **b** $\overline{K^0} \rightarrow \overline{K^0}$

In addition to the cross-transitions shown in Fig. 9.1, there are also self-transition as shown in Fig. 9.2. The size of the transition amplitudes in Figs. 9.1, 9.2 are all the same,

$$A_{KK} = A_{\overline{K}\overline{K}} = A_{K\overline{K}} = A_{\overline{K}K} \propto g_W^4 \sin^2 2\theta_C \ . \tag{9.3}$$

We use A_{KK} to represent all the transition amplitudes.

Using the transition amplitudes, the transition equation of the neutral K system is

$$i\frac{d}{dt}\begin{pmatrix} C_K \\ C_{\overline{K}} \end{pmatrix} = \begin{pmatrix} m_K + A_{KK} & A_{KK} \\ A_{KK} & m_K + A_{KK} \end{pmatrix}\begin{pmatrix} C_K \\ C_{\overline{K}} \end{pmatrix}, \tag{9.4}$$

where m_K is the original neutral K mass which comes from original d, s quark masses and strong potentials. K^0 and $\overline{K^0}$ have the same original mass due to the CPT theorem.

From the solution of the transition equation, the mass eigenstate becomes

$$\begin{cases} |\psi_+[t]\rangle = \frac{1}{\sqrt{2}}\left(|K^0\rangle + |\overline{K^0}\rangle\right)e^{-i(m_K+2A_{KK})t} \equiv |K_+^0\rangle e^{-i(m_K+2A_{KK})t} \\ |\psi_-[t]\rangle = \frac{1}{\sqrt{2}}\left(|K^0\rangle - |\overline{K^0}\rangle\right)e^{-im_Kt} \equiv |K_-^0\rangle e^{-im_Kt}. \end{cases} \tag{9.5}$$

The mass difference of the two mass eigenstates is

$$\Delta m_K = 2A_{KK}. \tag{9.6}$$

The general wave function of the neutral K system, Eq. (9.2) can also be expressed by a superposition of the mass eigenstate as

$$|\psi_K[t]\rangle = C_+ |K_+^0\rangle e^{-i(m_K+2A_{KK})t} + C_- |K_-^0\rangle e^{-im_Kt}. \tag{9.7}$$

If pure K^0 is produced at $t = 0$, the initial condition is

$$|\psi_K[0]\rangle = |K^0\rangle = C_+ |K_+^0\rangle + C_- |K_-^0\rangle = \frac{1}{\sqrt{2}}\left((C_+ + C_-)|K^0\rangle + (C_+ - C_-)|\overline{K^0}\rangle\right) \tag{9.8}$$

and $C_\pm$ are determined as

$$C_+ = C_- = \frac{1}{\sqrt{2}}. \tag{9.9}$$

The wave function at any time t is then

$$\begin{aligned} |\psi_K[t]\rangle &= \frac{|K_+^0\rangle e^{-iA_{KK}t} + |K_-^0\rangle e^{iA_{KK}t}}{\sqrt{2}} e^{-i(m_K+A_{KK})t} \\ &= \left(\cos A_{KK}t |K^0\rangle + i\sin A_{KK}t |\overline{K^0}\rangle\right)e^{-i(m_K+A_{KK})t}. \end{aligned} \tag{9.10}$$

The probability to find $\overline{K^0}$ at time t is

$$P[K^0 \rightarrow \overline{K^0} : t] = \left|\langle \overline{K^0}|\psi_K[t]\rangle\right|^2 = \sin^2 A_{KK}t. \tag{9.11}$$

This is the K^0-$\overline{K^0}$ oscillation.

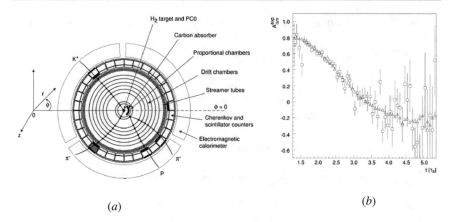

(a)   (b)

Fig. 9.3 **a** Transverse view of the CPLEAR detector. **b** Proper time distribution of the measured asymmetry $A_{\Delta m_K}$ (squares). The triangles are simulated asymmetries. The unit of the horizontal axis is K_S lifetime $\tau_S \sim 9.0 \times 10^{-11}$ s. From [1]

The transition amplitude A_{KK} can be measured from the oscillation frequency. A Δm_K measurement was performed by the CPLEAR collaboration [1]. The experimental apparatus is shown in Fig. 9.3a. They hit hydrogen gas with an antiproton beam and produced K^0 and $\overline{K^0}$ via the following strong interaction reactions:

$$\overline{p} + p \rightarrow \begin{cases} K^- + \pi^+ + K^0 \\ K^+ + \pi^- + \overline{K^0} . \end{cases} \tag{9.12}$$

The K^0 and $\overline{K^0}$ were separately tagged by the charge of the associated $K^\pm$. These neutral K's were detected after traveling proper time τ by the following strong interaction reactions with the proton or neutron in the carbon target:

$$\begin{cases} K^0 + p \rightarrow K^+ + n \\ \overline{K^0} + n \rightarrow K^- + p, \quad \overline{K^0} + n \rightarrow \pi^0 + \Lambda . \end{cases} \tag{9.13}$$

The Λ baryon decays to $p + \pi^-$ after traveling an order of 10 cm in the experimental apparatus and can easily be identified.

If the lifetime of neutral K is infinitely long, the number of events to detect for the combination of each production and detection mode are

$$n_{K^0 \rightarrow K^0}[\tau] = \mathcal{L} \epsilon \sigma \cos^2 \left[\frac{\Delta m_K}{2} \tau \right], \quad n_{K^0 \rightarrow \overline{K^0}}[\tau] = \mathcal{L} \epsilon \overline{\sigma} \sin^2 \left[\frac{\Delta m_K}{2} \tau \right],$$

$$n_{\overline{K^0} \rightarrow K^0}[\tau] = \mathcal{L} \overline{\epsilon} \sigma \sin^2 \left[\frac{\Delta m_K}{2} \tau \right], \quad n_{\overline{K^0} \rightarrow \overline{K^0}}[\tau] = \mathcal{L} \overline{\epsilon} \overline{\sigma} \cos^2 \left[\frac{\Delta m_K}{2} \tau \right],$$

$$\tag{9.14}$$

where ϵ and $\overline{\epsilon}$ are the tagging efficiency for generated K^0 and $\overline{K^0}$, and σ and $\overline{\sigma}$ are the product of the reaction cross section and detection efficiency for K^0 and $\overline{K^0}$,

Fig. 9.4 The transition amplitude A_{KK} can be calculated by integrating the diagrams with the possible 4-momentum k

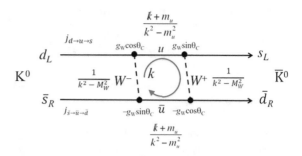

respectively. $\mathcal{L}$ is a common factor such as beam luminosity, target density, etc. If we measure the proper time distribution of the asymmetry defined below

$$A_{\Delta m_K}[\tau] = \frac{\left(n_{K\to K}/n_{\overline{K}\to K}\right) - \left(n_{K\to\overline{K}}/n_{\overline{K}\to\overline{K}}\right)}{\left(n_{K\to K}/n_{\overline{K}\to K}\right) + \left(n_{K\to\overline{K}}/n_{\overline{K}\to\overline{K}}\right)} = \frac{2\cos[\Delta m_K \tau]}{1 + \cos^2[\Delta m_K \tau]}, \quad (9.15)$$

the mass difference Δm_K can be measured free from the ambiguities of the experimental conditions. Figure 9.3b shows the data of $A_{\Delta m_K}[\tau]$.[1] From these data, the mass difference was measured as

$$\Delta m_K = 2A_{KK} = 3.6\,\mu\text{eV} \sim \frac{1}{1.9 \times 10^{-10}\,\text{s}}. \quad (9.16)$$

Now let's estimate the transition amplitude A_{KK} theoretically and compare it to the observation. The transition amplitude can be calculated by integrating the amplitudes of the diagrams in Fig. 9.1 with possible four-momentum k as shown in Fig. 9.4.

The mathematical formula is[2]

$$A_{K\overline{K}} = \frac{-ig_W^4 s_C^2 c_C^2}{4} \sum_{u_s, u_d, v_s, v_d} \int \frac{d^4k}{(2\pi)^4} \left(\frac{1}{k^2 - m_W^2}\right)^2 j_{d\to u\to s}^{\alpha\beta}[k] j_{\beta\alpha}^{\overline{s}\to\overline{u}\to\overline{d}}[k], \quad (9.17)$$

where $s_C = \sin\theta_C$, $c_C = \cos\theta_C$. The quark currents are

$$j_{d\to u\to s}^{\alpha\beta}[k] = \left\{\overline{u_s}\gamma^\alpha\gamma_L\left(\frac{\slashed{k} + m_u}{k^2 - m_u}\right)\gamma_L\gamma^\beta u_d\right\},$$

$$j_{\beta\alpha}^{\overline{s}\to\overline{u}\to\overline{s}}[k] = \left\{\overline{v_s}\gamma_\beta\gamma_L\left(\frac{\slashed{k} + m_u}{k^2 - m_u}\right)\gamma_\alpha\gamma_L v_d\right\}, \quad (9.18)$$

where u_q and v_q are 4 component spinors of positive and negative energy states and $\sum_{u_d, u_s, v_d, v_s}$ represents the sum of the initial and final state quark spins.

---

[1] Due to the decay effect, the actual formula of $A_{\Delta m_K}[\tau]$ is more complicate.
[2] Equation (7.24) of [2].

The result of the calculation predicts that the transition amplitude is [2]

$$A_{K\overline{K}} \sim \frac{1}{32\pi^2} G_F^2 m_W^2 f_K^2 m_K \sin^2 2\theta_C , \tag{9.19}$$

where G_F is the Fermi constant and f_K is the decay constant of K meson. Using the observed values,

$$\begin{aligned} G_F &\sim 1.16 \times 10^{-5} /\text{GeV}^2, \quad m_W \sim 80\,\text{GeV}, \\ f_K &\sim 0.16\,\text{GeV}, \quad m_K \sim 0.50\,\text{GeV}, \quad \theta_C \sim 0.23, \end{aligned} \tag{9.20}$$

the transition amplitude is calculated as

$$A_{K\overline{K}}^{\text{calc}} \sim 7.4\,\text{meV} . \tag{9.21}$$

However, this is 4,000 times larger than the observation (9.16). Historically, the discrepancy led to a motivation to introduce a new charge $+2/3$ quark, now called charm (c) quark. The u and c quarks couple to the flavor eigenstate d' and s' through the $W^\pm$ bosons. As described in Chap. 8, the weak eigenstate d' and s' are superposition of the mass eigenstate d and s, as follows:

$$\begin{pmatrix} s' \\ d' \end{pmatrix} = \begin{pmatrix} \cos\theta_C & -\sin\theta_C \\ \sin\theta_C & \cos\theta_C \end{pmatrix} \begin{pmatrix} s \\ d \end{pmatrix} \tag{9.22}$$

From this mixing, the quark-W boson couplings becomes

$$\begin{cases} g_W\{\bar{u}\gamma^\mu d'\}W_\mu = g_W \sin\theta_C\{\bar{u}\gamma^\mu s\}W_\mu + g_W \cos\theta_C\{\bar{u}\gamma^\mu d\}W_\mu \\ g_W\{\bar{c}\gamma^\mu s'\}W_\mu = g_W \cos\theta_C\{\bar{c}\gamma^\mu s\}W_\mu - g_W \sin\theta_C\{\bar{c}\gamma^\mu d\}W_\mu . \end{cases} \tag{9.23}$$

The introduction of c quark leads three more diagrams of $K^0 \to \overline{K^0}$ transitions as shown in Fig. 9.5. The amplitudes, $A_{K\overline{K}}^{cu}$ and $A_{K\overline{K}}^{uc}$ have opposite contribution with respect to $A_{K\overline{K}}^{uu}$ and $A_{K\overline{K}}^{cc}$ due to the structure of the mixing (9.22). We have to add all the diagrams to calculate the oscillation probability and the integrand of the quark propagator part of Eq. (9.17) is changed as

$$\begin{aligned} A_{K\overline{K}}^{uu} &+ A_{K\overline{K}}^{cc} + A_{K\overline{K}}^{cu} + A_{K\overline{K}}^{uc} \\ &\to \frac{k_\mu k^\nu}{(k^2 - m_W^2)^2} \left(\frac{1}{(k^2 - m_u^2)^2} + \frac{1}{(k^2 - m_c^2)^2} - \frac{2}{(k^2 - m_u^2)(k^2 - m_c^2)} \right) \\ &= \frac{(m_c^2 - m_u^2)^2 k_\mu k^\nu}{(k^2 - m_W^2)^2 (k^2 - m_c^2)^2 (k^2 - m_u^2)^2} . \end{aligned} \tag{9.24}$$

If $m_c = m_u$, the u-c mixed amplitude cancels u-u and c-c amplitudes and there is no transition between K^0 and $\overline{K^0}$. Contrary, the difference of c and u quarks masses can be obtained by comparing with the observed value.

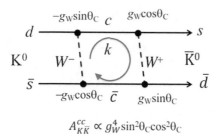

$$A_{K\overline{K}}^{cc} \propto g_W^4 \sin^2\theta_C \cos^2\theta_C$$

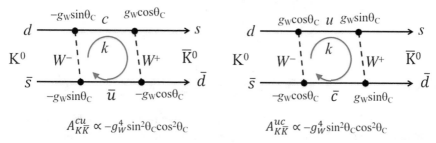

$$A_{K\overline{K}}^{cu} \propto -g_W^4 \sin^2\theta_C \cos^2\theta_C \qquad\qquad A_{K\overline{K}}^{uc} \propto -g_W^4 \sin^2\theta_C \cos^2\theta_C$$

Fig. 9.5 $K^0 \to \overline{K^0}$ diagram related to c quark. Note that $A_{K\overline{K}}^{cu}$ and $A_{K\overline{K}}^{uc}$ have opposite sign with respect to $A_{K\overline{K}}^{cc}$ and $A_{K\overline{K}}^{uu}$

The total $K^0 \to \overline{K^0}$ transition amplitude is obtained from the integration with possible 4-momentum k,

$$A_{K\overline{K}} = -4i G_F^2 m_W^4 \sin^2 2\theta_C \int \frac{d^4 k}{(2\pi)^4} \frac{(m_c^2 - m_u^2)^2 k_\mu k^\nu}{(k^2 - m_W^2)^2 (k^2 - m_c^2)^2 (k^2 - m_u^2)^2} T_\nu^\mu ,$$

(9.25)

where

$$T_\nu^\mu = \frac{1}{4} \sum_{u_s, u_d, v_s, v_d} \{\overline{u_s}\gamma^\alpha \gamma_L \gamma^\mu \gamma^\beta \gamma_L u_d\}\{\overline{v_s}\gamma_\beta \gamma_L \gamma_\nu \gamma_\alpha \gamma_L v_d\} .$$

(9.26)

Under the condition $m_W \gg m_c \gg m_u$, the integration results in the transition amplitude,[3]

$$A_{K\to\overline{K}} \sim \frac{1}{32\pi^2} G_F^2 m_c^2 f_K^2 m_K \sin^2 2\theta_C \sim 1.1 \times 10^{-15} \, m_c^2 / \mathrm{GeV}^2 .$$

(9.27)

Using the measured value (9.16), the c-quark mass is predicted as

$$m_c \sim 1.3 \, \mathrm{GeV} .$$

(9.28)

This value agrees with the PDG value; $m_c = 1.27 \pm 0.02 \, \mathrm{GeV}$.

---

[3]p. 254 of [2].

9.3   Six-Quark System and CP Violation

In the previous section, only u and c quarks are taken into consideration. However, there is another heavy ($m \sim 170\,\text{GeV}$), charge $+2/3$ quark called top quark. If we include the top quark in the system, the $K^0 - \overline{K^0}$ transitions become as shown in Fig. 9.6.

The transition amplitude is expressed as [2]

$$
\begin{aligned}
A_{K\overline{K}} = {} & \frac{g_W^4}{4} \sum_{i,j} V_{id} V_{is}^* V_{js}^* V_{jd} \\
& \times \sum_{u_d, u_s, v_d, v_s} \{\overline{v_s}\gamma_\rho\gamma_L\gamma_\alpha\gamma_\nu\gamma_L v_d\}\{\overline{u_s}\gamma^\nu\gamma_L\gamma^\beta\gamma^\rho\gamma_L u_d\} \\
& \times \int \frac{d^4k}{(2\pi)^4} \frac{k^\alpha k_\beta}{(k^2 - m_W^2)^2(k^2 - m_j^2)(k^2 - m_i^2)} \; .
\end{aligned}
\tag{9.29}
$$

A difference from (9.17) is that CKM matrix elements, V_{ij} are used instead of the Cabibbo angle θ_C. The values of the CKM matrix elements are shown in Eq. (8.61). There is an important difference from the two-flavor case, that V_{ij} include an imaginary number which leads CP violation. In this case, the transition amplitude can be parametrized as

$$
A_{K\overline{K}} = K g_W^4 \left(\sum_{i=u,c,t} V_{id} V_{is}^* \Pi[m_i] \right)^2 \equiv A \exp[+i\alpha] \, ,
\tag{9.30}
$$

where $\Pi[m_i]$ represents the propagator of quark-i. Since the imaginary component comes in as very small V_{td}, α is assumed to be very small, too. The reverse transition is shown in Fig. 9.7, which is the complex conjugate of $A_{K\overline{K}}$,

$$
A_{\overline{K}K} = K g_W^4 \left(\sum_{i=u,c,t} V_{id}^* V_{is} \Pi[m_i] \right)^2 = A \exp[-i\alpha] = A_{K\overline{K}}^* \; .
\tag{9.31}
$$

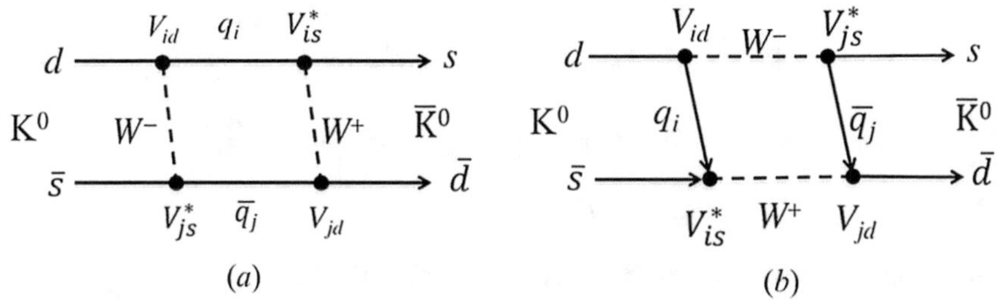

Fig. 9.6 $K^0 \to \overline{K^0}$ transition. Intermediate q_i, q_j are u, c or t quark. The transition amplitude is parametrized $Ae^{i\alpha}$

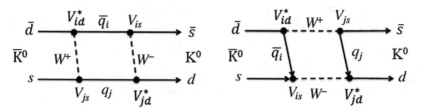

Fig. 9.7 $\overline{K^0} \rightarrow K^0$ transition. Intermediate q_i, q_j are u, c and t quarks. The transition amplitude is parametrized as $Ae^{-i\alpha}$

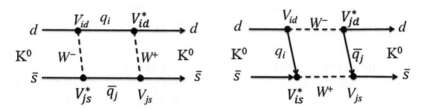

Fig. 9.8 $K^0 \rightarrow K^0$ transition. Intermediate q_i, q_j are u, c and t quarks. The transition amplitude is A_{KK}

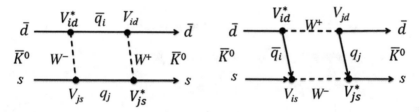

Fig. 9.9 $\overline{K^0} \rightarrow \overline{K^0}$ transition. Intermediate q_i, q_j are u, c and t quarks. The transition amplitude is A

There are also $K^0 \rightarrow K^0$ and $\overline{K^0} \rightarrow \overline{K^0}$ self-transitions as shown in Fig. 9.8 and Fig. 9.9, respectively.   For these transitions, the transition amplitudes are real,

$$A_{KK} = A_{\overline{K}\overline{K}} = K g_W^4 \left| \sum_{i=u,c,t} V_{id} V_{is}^* \Pi[m_i] \right|^2 = A . \tag{9.32}$$

Figure 9.10 summarizes the transition diagram for K^0-$\overline{K^0}$ system for the six-quark system.   Due to the transition, the coefficients of the superposition (9.2) satisfy the following transition equation:

$$i\frac{d}{dt}\begin{pmatrix} C_K \\ C_{\overline{K}} \end{pmatrix} = \left(M_K + A_{KK}\begin{pmatrix} 1 & e^{-i\alpha} \\ e^{i\alpha} & 1 \end{pmatrix} \right)\begin{pmatrix} C_K \\ C_{\overline{K}} \end{pmatrix} , \tag{9.33}$$

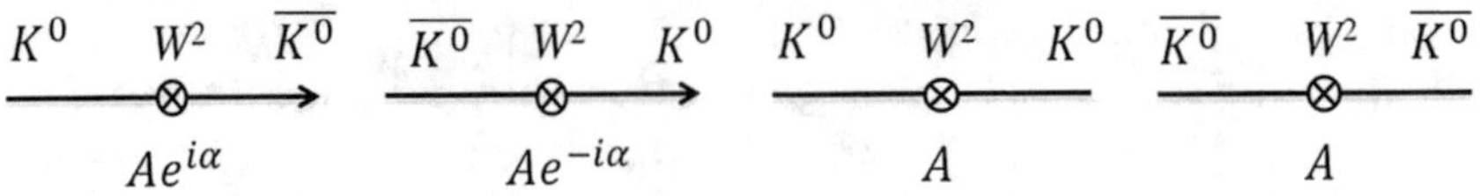

$$K^0 \quad W^2 \quad \overline{K^0} \qquad \overline{K^0} \quad W^2 \quad K^0 \qquad K^0 \quad W^2 \quad K^0 \qquad \overline{K^0} \quad W^2 \quad \overline{K^0}$$

$$\xrightarrow{\quad\otimes\quad} \qquad \xrightarrow{\quad\otimes\quad} \qquad \xrightarrow{\quad\otimes\quad} \qquad \xrightarrow{\quad\otimes\quad}$$

$$Ae^{i\alpha} \qquad\qquad Ae^{-i\alpha} \qquad\qquad A \qquad\qquad A$$

Fig. 9.10 $(K^0 \leftrightarrow \overline{K^0})$, $(\overline{K^0} \leftrightarrow K^0)$ and $(\overline{K^0} \leftrightarrow \overline{K^0})$ transition amplitudes. W^2 indicates these transitions are caused by the second-order weak interactions

where $M_K = \begin{pmatrix} m_K & 0 \\ 0 & m_K \end{pmatrix}$. From this, the mass eigenstate is

$$\begin{cases} |\psi_+[t]\rangle = \frac{1}{\sqrt{2}} \left(|K^0\rangle + e^{-i\alpha}|\overline{K^0}\rangle \right) e^{-i(m_K+2A)t} \\ |\psi_-[t]\rangle = \frac{1}{\sqrt{2}} \left(-e^{i\alpha}|K^0\rangle + |\overline{K^0}\rangle \right) e^{-im_K t} . \end{cases} \tag{9.34}$$

The important difference from Eq. (9.5) is that the amplitude of the superposition is a complex number.

The general wave function can be written as a superposition of the mass eigenstates,

$$|\Psi[t]\rangle = C_+ |\psi_+[t]\rangle + C_- |\psi_-[t]\rangle . \tag{9.35}$$

9.3.1 $K^0 - \overline{K^0}$ Oscillation of Six-Quark System

If the system is pure K^0 at $t = 0$, from (9.34) and (9.35),

$$\begin{aligned} |K^0\rangle = \Psi[0] &= C_+\psi_+[0] + C_-\psi_-[0] \\ &= \frac{1}{\sqrt{2}}(C_+ - e^{i\alpha}C_-)|K^0\rangle + \frac{1}{\sqrt{2}}(e^{-i\alpha}C_+ + C_-)|\overline{K^0}\rangle . \end{aligned} \tag{9.36}$$

Then we obtain

$$C_+ = \frac{1}{\sqrt{2}}, \quad C_- = -\frac{1}{\sqrt{2}}e^{-i\alpha} . \tag{9.37}$$

The wave function at time t is

$$\Psi[t] = \frac{1}{2}e^{-im_K t}\left(\left(1 + e^{-2iAt}\right)|K^0\rangle - e^{-i\alpha}\left(1 - e^{-2iAt}\right)|\overline{K^0}\rangle \right) . \tag{9.38}$$

Therefore, the oscillation probability is

$$P_{K^0 \to \overline{K^0}}[t] = \left| \frac{1}{2}e^{-i\alpha}(e^{-2iAt} - 1) \right|^2 = \sin^2[At] . \tag{9.39}$$

Note that this $K^0 \leftrightarrow \overline{K^0}$ oscillation probability does not depend on α. This means that CP violation effect does not directly emerge to the $K^0 \leftrightarrow \overline{K^0}$ oscillation probability. The decay of K^0 and $\overline{K^0}$ plays an important role in the CP violation effect.

9.3.2 Oscillation of K^0 CP Eigenstate

The same discussion in the previous section can be made on CP eigenstate bases. The CP operation transforms neutral K system as follows:

$$\begin{cases} CP\left|K^0\right\rangle = -C\left|K^0\right\rangle = -\left|\overline{K^0}\right\rangle \\ CP\left|\overline{K^0}\right\rangle = -C\left|\overline{K^0}\right\rangle = -\left|K^0\right\rangle \,, \end{cases} \tag{9.40}$$

where, the fact that the meson parity is negative is used. From this, the following relation can be obtained:

$$CP\left(\left|K^0\right\rangle \pm \left|\overline{K^0}\right\rangle\right) = \mp\left(\left|K^0\right\rangle \pm \left|\overline{K^0}\right\rangle\right) \quad \text{(double} - \text{sign corresponds)} \tag{9.41}$$

Therefore, the CP eigenstates of the K^0-$\overline{K^0}$ system are

$$\begin{pmatrix} \left|K_+^{CP}\right\rangle \\ \left|K_-^{CP}\right\rangle \end{pmatrix} = \frac{1}{\sqrt{2}} \begin{pmatrix} 1 & -1 \\ 1 & 1 \end{pmatrix} \begin{pmatrix} \left|K^0\right\rangle \\ \left|\overline{K^0}\right\rangle \end{pmatrix} . \tag{9.42}$$

The general wave function can be equally expressed by superpositions of K^0-$\overline{K^0}$ base and $K_\pm^{CP}$ base as

$$\psi_K[t] = C_K[t]\left|K^0\right\rangle + C_{\overline{K}}[t]\left|\overline{K^0}\right\rangle = C_+^{CP}[t]\left|K_+^{CP}\right\rangle + C_-^{CP}[t]\left|K_-^{CP}\right\rangle . \tag{9.43}$$

By comparing both base states using the relation (9.42), we obtain the relation of the coefficients as

$$\begin{pmatrix} C_+^{CP} \\ C_-^{CP} \end{pmatrix} = \begin{pmatrix} \langle K_+^{CP}|\psi_K\rangle \\ \langle K_-^{CP}|\psi_K\rangle \end{pmatrix} = \frac{1}{\sqrt{2}} \begin{pmatrix} 1 & -1 \\ 1 & 1 \end{pmatrix} \begin{pmatrix} C_K \\ C_{\overline{K}} \end{pmatrix} . \tag{9.44}$$

Note that this is exactly the same relation between the flavor and CP basis vectors (9.42).

The transition equation for CP basis is, by combining Eqs. (9.33) and (9.44),

$$i\frac{d}{dt}\begin{pmatrix} C_-^{CP} \\ C_+^{CP} \end{pmatrix} = \left(M_K + A \begin{pmatrix} 2\cos^2[\alpha/2] & i\sin\alpha \\ -i\sin\alpha & 2\sin^2[\alpha/2] \end{pmatrix} \right) \begin{pmatrix} C_-^{CP} \\ C_+^{CP} \end{pmatrix} . \tag{9.45}$$

The transition diagram is shown in Fig. 9.11 and the mixing triangle is shown in Fig. 9.12. At CP bases, the mass eigenstates are

$$\begin{cases} |\psi_+[t]\rangle = (\sin[\alpha/2]\left|K_+^{CP}\right\rangle + i\cos[\alpha/2]\left|K_-^{CP}\right\rangle)e^{-i(m_K+2A)t} \\ |\psi_-[t]\rangle = (\cos[\alpha/2]\left|K_+^{CP}\right\rangle - i\sin[\alpha/2]\left|K_-^{CP}\right\rangle)e^{-im_K t} \,. \end{cases} \tag{9.46}$$

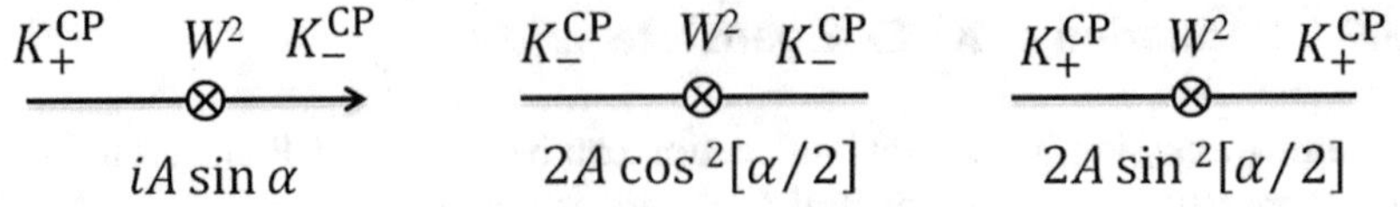

$$\underset{iA \sin \alpha}{\underset{\otimes}{\underrightarrow{K^{CP}_+ \qquad W^2 \qquad K^{CP}_-}}} \qquad \underset{2A \cos^2[\alpha/2]}{\underset{\otimes}{\underline{K^{CP}_- \qquad W^2 \qquad K^{CP}_-}}} \qquad \underset{2A \sin^2[\alpha/2]}{\underset{\otimes}{\underleftarrow{K^{CP}_+ \qquad W^2 \qquad K^{CP}_+}}}$$

Fig. 9.11 K^{CP} transition amplitudes

Fig. 9.12 K^{CP} mixing triangle

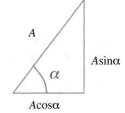

This means if $\alpha = 0$, the CP eigenstate and mass eigenstate coincide and the CP does not change in time. However, if $\alpha \neq 0$, The $K^{CP}_+ \leftrightarrow K^{CP}_-$ oscillation takes place as follows.

$$P_{K^{CP}_+ \leftrightarrow K^{CP}_-}[t] = \sin^2 \alpha \sin^2[At] \tag{9.47}$$

This means that if α is finite, the CP eigenvalue does not conserve. This is the origin of the indirect CP violation in K^0-$\overline{K^0}$ system.

9.4   Discovery of CP Violation and Measurement of α

The K^0 meson can decay to 2π system. Since the K^0 spin is 0, the angular momentum between the two π mesons is $l = 0$ and the CP quantum number of the 2π system is CP$= +1$.[4] Therefore, if CP is conserved in the decay process of the K^0 meson, $K^{CP}_- \to 2\pi$ is forbidden, while $K^{CP}_+ \to 2\pi$ decay is allowed. Therefore, the K^{CP}_+ component of the K^0 system can be measured from their 2π decays. On the other hand, the main decay modes of K^{CP}_- are 3π and $\pi l \nu$. They are both 3 body decays and suppressed much compared with the 2π decays. This means the lifetime of K^{CP}_- component is much longer than K^{CP}_+.

Experimentally, K^0 are observed to have a short and long-lifetime components, called K_S (K-Short) and K_L (K-Long), respectively. Their lifetimes are

$$\tau_S \sim 9.0 \times 10^{-11}\,\text{s}\,, \quad \tau_L \sim 5.1 \times 10^{-8}\,\text{s}\,. \tag{9.48}$$

---

[4]Because $\pi^\pm$ are bosons, the wave function of the system is $\psi_{2\pi} = (|\pi^+\pi^-\rangle + |\pi^-\pi^+\rangle)/\sqrt{2}$ due to the Pauli exchange principle. Then the CP eigenvalue of this system is CP$\psi_{2\pi} = (|(-\pi^-)(-\pi^+)\rangle + |(-\pi^+)(-\pi^-)\rangle)/\sqrt{2} = +\psi_{2\pi}$.

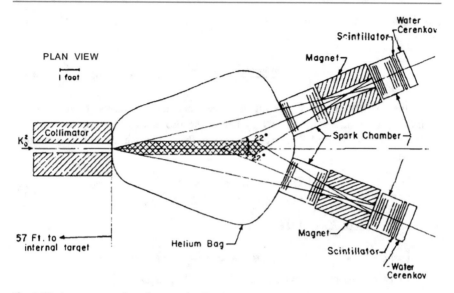

Fig. 9.13 Apparatus used to discover the CP violation. K_L came in from the left and decays in the He bag. A pair of spectrometer measures the momentum of charged particles generated in the decay. From [3] ($^{\odot}$The Nobel Foundation)

From the above discussions it can be assumed that

$$K_L = K_-^{CP}, \qquad K_S = K_+^{CP}. \tag{9.49}$$

In 1960s, the time dependence of the $K^0 \rightarrow \pi^+ \pi^-$ decay branching ratio was measured. The experimental group of J. Cronin and V.I. Fitch produced K^0 by strong interactions,

$$p + A \rightarrow K^0 + \Lambda + X, \tag{9.50}$$

at the AGS accelerator. From (9.42), the generated K^0 is a superposition of $K_\pm^{CP}$ that are regarded as K_S and K_L,

$$\left|K^0\right\rangle = \frac{1}{\sqrt{2}}(\left|K_+^{CP}\right\rangle + \left|K_-^{CP}\right\rangle) \rightarrow \frac{1}{\sqrt{2}}(\left|K_S\right\rangle + \left|K_L\right\rangle). \tag{9.51}$$

Since $c\tau_S \sim 2.7$ cm, K_S decays out after traveling a few meters and only K_L component remains after that. The group tried to measure $K_L \rightarrow \pi^+ \pi^-$ decay events using a pair of spectrometers shown in Fig. 9.13. Surprisingly they found 0.2% of K_L decay to $\pi^+ \pi^-$. From the relation (9.49), it seems that CP $= -$ state decays to CP $= +$ state and the decay seems to violate the CP symmetry. This is the discovery of the CP violation.

What is happening here is as follows. When K^0 was produced by the strong interaction (9.50), from the relation (9.51), 50% was K_+^{CP} and another 50% was K_-^{CP}. K_+^{CP} decays quickly to $\pi^+ \pi^-$ and disappears. On the other hand, most of K_-^{CP}

remains and a small part of K_-^{CP} turns to K_+^{CP} due to the oscillation (9.47) and the generated K_+^{CP} decays to $\pi^+\pi^-$.

$$K_-^{\mathrm{CP}} \xrightarrow{\quad \text{Oscillation} \quad} K_+^{\mathrm{CP}} \xrightarrow{\quad \text{Decay} \quad} \pi^+\pi^- \tag{9.52}$$

Since the lifetime of the K_L is much longer than the oscillation period and the oscillation amplitude $(\sin^2\alpha)$ is very small, the process can be approximated as stationary and the K_L and K_S can be regarded as the superpositions of CP eigenstates as shown in (9.46),

$$\begin{cases} |K_L\rangle \sim (\alpha/2)\,|K_+^{\mathrm{CP}}\rangle + i\,|K_-^{\mathrm{CP}}\rangle \\ |K_S\rangle \sim |K_+^{\mathrm{CP}}\rangle - i(\alpha/2)\,|K_-^{\mathrm{CP}}\rangle \;. \end{cases} \tag{9.53}$$

The decay width of K_L and K_S to $\pi\pi$ is

$$\begin{cases} \Gamma[K_S \to \pi\pi] \propto |\langle\pi\pi|H_W|K_S\rangle|^2 \sim |\langle\pi\pi|H_W|K_+^{\mathrm{CP}}\rangle|^2 \\ \Gamma[K_L \to \pi\pi] \propto |\langle\pi\pi|H_W|K_L\rangle|^2 \sim |\alpha/2|^2|\langle\pi\pi|H_W|K_+^{\mathrm{CP}}\rangle|^2 \;. \end{cases} \tag{9.54}$$

From the measured lifetime and branching ratios,

$$|\alpha| = 2\sqrt{\frac{\Gamma[K_L \to \pi\pi]}{\Gamma[K_S \to \pi\pi]}} = 2\sqrt{\frac{\Gamma[K_L \to \pi\pi]}{\Gamma[K_L \to all]}\frac{\tau_S}{\tau_L}} \sim 4.43 \times 10^{-3} \tag{9.55}$$

is observed.

The decay in Eq. (9.52) is actually from CP $= +$ state to CP $= +$ state and CP is not violated in the decay process. The origin of the CP non conservation is the oscillation of the CP eigenstate ($K_-^{\mathrm{CP}} \to K_+^{\mathrm{CP}}$). Therefore, this type of CP violation is called indirect CP violation. Later, CP violation in the decay process, called direct CP violation was also discovered, which can also be explained by the same imaginary component of the CKM matrix.

References

1. Angelopoulos, A., et al.: Phys. Lett. B **503**, 49 (2001)
2. Commins, E.D., Buckbaus, P.H.: Weak Interactions of Leptons and Quarks. Cambridge University Press, Cambridge (1983)
3. Fitch, V.L.: Nobel Lecture (1989). https://www.nobelprize.org/uploads/2018/06/fitch-lecture.pdf

Part IV
Strong Interactions

Quark Structure of Mesons

10

10.1 Introduction

Mesons are binding system of a quark and an anti-quark by the strong interactions. For example, $\left|\pi^+\right\rangle = \left|u\bar{d}\right\rangle$, $\left|\pi^0\right\rangle = (\left|u\bar{u}\right\rangle - \left|d\bar{d}\right\rangle)/\sqrt{2}$, $\left|K^+\right\rangle = \left|u\bar{s}\right\rangle$, $\left|\rho^0\right\rangle = (\left|u\bar{u}\right\rangle - \left|d\bar{d}\right\rangle)/\sqrt{2}$, etc. The quark structure of π^0 and ρ^0 are the same but their spins are different.

Table 10.1 shows quark and anti-quark combinations and corresponding mesons and their masses. Since the quark spin is 1/2, the spin of low mass ($l = 0$) mesons are either $S = 0$ or $S = 1$. As Table 10.1 shows, the masses of the mesons are very much different for different spin although the quark contents are the same. In some cases, there are more than one mesons in a box. This is because different $q\bar{q}$ states mix and some mesons contain more than one $q\bar{q}$ component. This mixing is also different for different spin states. Those properties can be understood from the mixing and oscillation based on the simple quark model as we see in the following sections.

10.2 u, d, s-Quark Masses

First, we obtain a crude sense of the light quark masses. Since quarks cannot exist alone, it is impossible to directly measure their masses. The hadron mass is the sum of the quark masses and the strong and electromagnetic potentials. It is difficult to separate the contribution of the quark masses and the potentials. In fact, there are two kinds of definitions of quark mass in elementary particle physics. One is called "current quark mass" which is used for ab initio calculation of QCD. Another one is "constituent quark mass" which is obtained from the hadron masses assuming the potential contributions are relatively small. In the following quark model of hadrons, the constituent quark masses are used. Although they are empirical parameters, many

© Springer Nature Switzerland AG 2021
F. Suekane, *Quantum Oscillations*, Lecture Notes in Physics 985,
https://doi.org/10.1007/978-3-030-70527-5_10

Table 10.1 Quark and anti-quark combinations and corresponding mesons. The values in parenthesis are the masses in MeV/c^2. Strictly speaking, K^0 and $\overline{K^0}$ are not mass eigenstate (See Chap. 9). We call Spin = 0 mesons *Pseudoscalar* meson because their quantum number is $J^P = 0^-$ and we call Spin = 1 mesons *Vector* meson because their quantum number is $J^P = 1^-$

Spin=0	u	d	s
$\overline{u}$	$\pi^0(135)$, $\eta(548)$, $\eta'(958)$	$\pi^-(140)$	$K^-(494)$
$\overline{d}$	$\pi^+(140)$	π^0, η, η'	$\overline{K^0}(498)$
$\overline{s}$	$K^+(494)$	$K^0(498)$	η, η'

Spin=1	u	d	s
$\overline{u}$	$\rho^0(775)$, $\omega(783)$	$\rho^-(775)$	$K^{-*}(892)$
$\overline{d}$	$\rho^+(775)$	ρ^0, ω	$\overline{K^{0*}}(896)$
$\overline{s}$	$K^{+*}(892)$	$K^{0*}(896)$	$\phi(1019)$

properties of hadrons can successfully be explained by using the constituent quark masses as we will see. In fact, it is still a mystery why the constituent quark mass can explain so successfully various properties of the hadrons although the current quark mass is considered to be the true theory.

Figure 10.1 shows the baryon mass dependence on the number of constituent strange quarks. Assuming the baryon masses are the sum of the constituent quark masses and a small potential energy, the following rough mass values are obtained from the mass pattern:

$$m_u \sim m_d = 300 \sim 400 \text{ MeV}, \quad m_s \sim m_q + 150 \text{ MeV} = 450 \sim 550 \text{ MeV} , \tag{10.1}$$

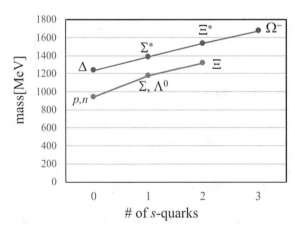

Fig. 10.1 Baryon mass and the number of constituent strange quarks. The baryons on the (p, n)-line have spin-1/2 and those on the Δ-line have spin-3/2. The masses show monotonous increase depending on the number of constituent s-quark. Historically, the mass of the Ω^- baryon had been predicted from this mass pattern and later it was discovered at the predicted mass region. The mass difference between spin-1/2 and spin-3/2 baryons comes from the spin-dependent strong potential

where q represents u or d quark. The difference between u and d-quark masses is

$$m_d - m_u \sim m_n - m_p = 1.3 \text{ MeV} . \tag{10.2}$$

Since the mass difference of u- and d-quarks is much smaller than their absolute masses, practically it is possible to treat the u and d-quarks equivalently. This is called isosymmetry. We will use

$$m_u = m_d \equiv m_q \sim 350 \text{ MeV}, \quad m_s \sim 500 \text{ MeV} \tag{10.3}$$

as reference values, unless otherwise specified.

10.3 π^+-ρ^+ Mass Difference

π^+ and ρ^+ mesons have the same quark contents $|u\overline{d}\rangle$. However, their masses are pretty much different (140 MeV and 775 MeV, respectively). What causes this large mass difference?

In those mesons, u and $\overline{d}$-quarks are bound with respect to each other exchanging gluons as shown in Fig. 10.2a. The strong coupling constant of quarks is g_S and that of anti-quarks is $-g_S$ and the transition amplitude A_S is proportional to $-g_S^2$. A_S corresponds to the strong potential and because the strong potential is negative, the gluon exchange causes an attractive force and u and $\overline{d}$ are bound in π^+ and ρ^+ mesons. This effect is represented as self-transition of $|u\overline{d}\rangle$ system as shown in Fig. 10.2b,

$$|u\overline{d}\rangle \xleftrightarrow{A_S} |u\overline{d}\rangle . \tag{10.4}$$

By expressing the wave function as

$$|\psi[t]\rangle = C[t] |u\overline{d}\rangle , \tag{10.5}$$

$$A_S \propto -g_S^2$$

$$(a) \qquad\qquad\qquad\qquad (b)$$

Fig. 10.2 **a** u and $\overline{d}$-quarks are bound exchanging gluons. In this figure, total gluon effects are represented by one exchange line. The strong coupling constant of quarks is g_S and that of anti-quarks is $-g_S$. The transition amplitude A_S is proportional to $-g_S^2$. A_S corresponds to the strong potential. Because the strong potential is negative, the gluon exchange causes an attractive force. **b** The self-transition amplitude of $|u\overline{d}\rangle$ system of the diagram (**a**). The quark masses (m_u, m_d) and the strong potential (A_S) contribute to the self-transition

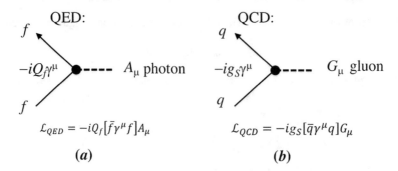

Fig. 10.3 a QED ferminon-fermion-photon vertex ($ff\gamma$). **b** QCD qqG vertex. Both Lagrangians have the same mathematical structure

the self-transition equation can be written as follows:

$$i\frac{d}{dt}C[t] = (m_u + m_d + A_S)C[t] = (2m_q + A_S)C[t].\tag{10.6}$$

The solution is

$$C[t] = C[0]e^{-i(2m_q+A_S)t}\tag{10.7}$$

and the wave function is

$$|\psi[t]\rangle = C[0]\,|u\overline{d}\rangle\,e^{-i(2m_q+A_S)t} = |\psi[0]\rangle\,e^{-i(2m_q+A_S)t}.\tag{10.8}$$

The mass of the u-$\overline{d}$ system is

$$m_{u\overline{d}} = 2m_q + A_S.\tag{10.9}$$

The structural difference between π^+ and ρ^+ is only the spin. The spin of π^+ is $S = 0$, while the spin of ρ^+ is $S = 1$. Therefore, the mass difference should be attributed to the spin dependence of the strong potential A_S.

In the standard model, the quark-gluon vertex and the fermion-photon vertex have the same structure as shown in Fig. 10.3.[1] The Lagrangian density of the QED and QCD vertexes are written as follows:

$$\mathcal{L}_{ff\gamma} = -iQ_f[\overline{f}\gamma_\mu f]A^\mu, \quad \mathcal{L}_{qqg} = -ig_S[\overline{q}\gamma_\mu q]G^\mu,\tag{10.10}$$

where A^μ and G^μ represent the photon and gluon fields. g_S is the gluon-quark coupling constant which is universal for all quarks. The strength of the strong coupling corresponding to the fine structure constant in QED is written as

$$\alpha_S = \frac{g_S^2}{4\pi}.\tag{10.11}$$

---

[1]The difference between the EM interactions and the strong interactions comes from the three gluon coupling. There are no three-photon couplings in QED.

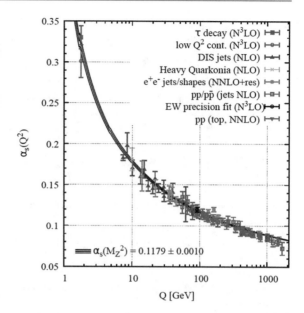

Fig. 10.4 Running α_S value. The binding energy of the quarks in hadrons is $Q = O[1\,\text{GeV}]$ and $\alpha_S \sim O[1]$. From [1]

The value of α_S depends on the momentum transfer Q as shown in Fig. 10.4 due to the self-coupling of the gluons. For interactions of quarks in hadrons, $Q = O[1\,\text{GeV}]$ and $\alpha_S = O[1]$.

Just as the electrostatic field (E) and the magnetic moment ($\vec{\mu}$) are derived from the QED Lagrangian as

$$\mathcal{L}_{ff\gamma} \longrightarrow E = \frac{e}{4\pi r^2}, \quad \vec{\mu} = \frac{e}{2m_f}\vec{\sigma}, \tag{10.12}$$

so the "strong" electrostatic field (E_S) and "strong" magnetic moment ($\vec{\mu}_S$) are derived from the QCD Lagrangian as,

$$\mathcal{L}_{qqg} \longrightarrow E_S = \frac{g_S}{4\pi r^2}, \quad \vec{\mu}_S = \frac{g_S}{2m_q}\vec{\sigma}. \tag{10.13}$$

If we use $\alpha_S \sim 1$ and $a_M \sim 1$ fm (approximate size of hadrons), the static strong potential between two quarks is very roughly estimated as

$$V_S = -\frac{\alpha_S}{a_M} \sim -200\,\text{MeV}. \tag{10.14}$$

This is a similar size to the constituent quark mass of light quarks (m_q).

As for the magnetic moment, there is magnetic moment-magnetic moment (MM-MM) interactions like in the hydrogen atom that generates the HI line. Classical energy shift due to MM-MM interaction is, from the analogy of Eq. (3.11),

$$E_M \propto (\vec{\mu}_u \cdot \vec{\mu}_{\bar{d}}). \tag{10.15}$$

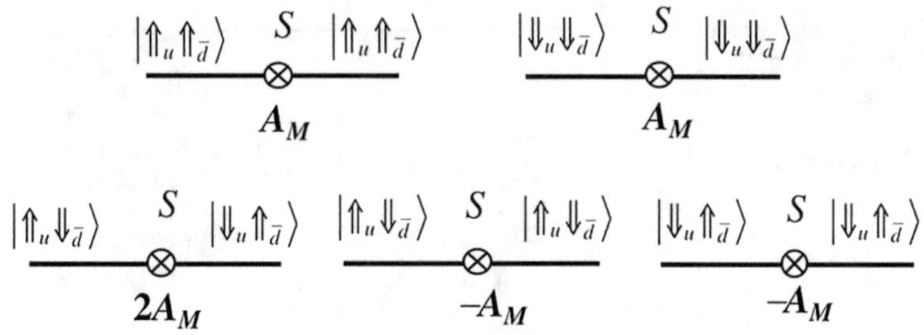

$$|\Uparrow_u \Uparrow_{\bar{d}}\rangle \xrightarrow[\otimes]{S} |\Uparrow_u \Uparrow_{\bar{d}}\rangle \qquad |\Downarrow_u \Downarrow_{\bar{d}}\rangle \xrightarrow[\otimes]{S} |\Downarrow_u \Downarrow_{\bar{d}}\rangle$$
$$A_M \qquad\qquad\qquad A_M$$

$$|\Uparrow_u \Downarrow_{\bar{d}}\rangle \xrightarrow[\otimes]{S} |\Downarrow_u \Uparrow_{\bar{d}}\rangle \quad |\Uparrow_u \Downarrow_{\bar{d}}\rangle \xrightarrow[\otimes]{S} |\Uparrow_u \Downarrow_{\bar{d}}\rangle \quad |\Downarrow_u \Uparrow_{\bar{d}}\rangle \xrightarrow[\otimes]{S} |\Downarrow_u \Uparrow_{\bar{d}}\rangle$$
$$2A_M \qquad\qquad -A_M \qquad\qquad -A_M$$

Fig. 10.5 u and $\bar{d}$ spin transitions due to the strong magnetic moment-magnetic moment (MM-MM) interactions

After the quantization, the state Hamiltonian of the MM-MM interactions becomes

$$H_M = -\frac{2\alpha_S}{3a_M^3 m_q^2}(\vec{\sigma}_u \cdot \vec{\sigma}_{\bar{d}}) \equiv A_M(\vec{\sigma}_u \cdot \vec{\sigma}_{\bar{d}}) . \tag{10.16}$$

This interaction is spin-dependent and is assumed to generate the difference of π^+ and ρ^+ masses.

The spin part of the wave function of $(u\text{-}\bar{d})$ system can be written as

$$\psi_{u\bar{d}}[t] = |u\bar{d}\rangle (C_{\Uparrow\Uparrow}[t]\,|\Uparrow\Uparrow\rangle + C_{\Uparrow\Downarrow}[t]\,|\Uparrow\Downarrow\rangle + C_{\Downarrow\Uparrow}[t]\,|\Downarrow\Uparrow\rangle + C_{\Downarrow\Downarrow}[t]\,|\Downarrow\Downarrow\rangle), \tag{10.17}$$

where the first arrow shows the spin of u-quark and the second arrow shows the spin of $\bar{d}$-quark. Following the discussions in Chap. 3, the transition amplitude between the spin combination is shown in Fig. 10.5 and the transition equation of C_{SS} is

$$i\frac{d}{dt}\begin{pmatrix} C_{\Uparrow\Uparrow} \\ C_{\Uparrow\Downarrow} \\ C_{\Downarrow\Uparrow} \\ C_{\Downarrow\Downarrow} \end{pmatrix} = A_M \begin{pmatrix} 1 & 0 & 0 & 0 \\ 0 & -1 & 2 & 0 \\ 0 & 2 & -1 & 0 \\ 0 & 0 & 0 & 1 \end{pmatrix} \begin{pmatrix} C_{\Uparrow\Uparrow} \\ C_{\Uparrow\Downarrow} \\ C_{\Downarrow\Uparrow} \\ C_{\Downarrow\Downarrow} \end{pmatrix} . \tag{10.18}$$

The mass eigenstates are

$$\psi_{\mathrm{I}} = |u\bar{d}\rangle\,|\Uparrow\Uparrow\rangle\, e^{-i(2m_q+V_S+A_M)t}, \quad \psi_{\mathrm{II}} = |u\bar{d}\rangle\,|\Downarrow\Downarrow\rangle\, e^{-i(2m_q+V_S+A_M)t}$$
$$\psi_{\mathrm{III}} = |u\bar{d}\rangle\,\frac{|\Uparrow\Downarrow\rangle + |\Downarrow\Uparrow\rangle}{\sqrt{2}}\, e^{-i(2m_q+V_S+A_M)t}, \tag{10.19}$$
$$\psi_{\mathrm{IV}} = |u\bar{d}\rangle\,\frac{|\Uparrow\Downarrow\rangle - |\Downarrow\Uparrow\rangle}{\sqrt{2}}\, e^{-i(2m_q+V_S-3A_M)t} .$$

The states $\psi_{\mathrm{I}} \sim \psi_{\mathrm{III}}$ have the same mass contributions; A_M, and ψ_{IV} has the different mass contribution $-3A_M$. Therefore, it is expected that ψ_{IV} corresponds to π^+ and

Fig. 10.6 Level structure of $|u\bar{d}\rangle$ system based on the simple quark model. V_S is *strong* electrostatic potential and A_M is the potential of the *strong* MM-MM interactions

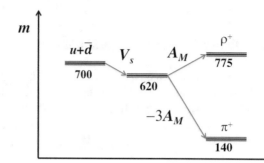

$\psi_{\mathrm{I}\sim\mathrm{III}}$ corresponds to the three spin states of ρ^+. Their wave function and total mass can be expressed as follows:

$$|\pi^+\rangle = |u\bar{d}\rangle\frac{|\uparrow\Uparrow\downarrow\Downarrow\rangle - |\downarrow\Downarrow\uparrow\Uparrow\rangle}{\sqrt{2}}; \quad m_\pi = 2m_q + V_S - 3A_M$$

$$|\rho^+\rangle = |u\bar{d}\rangle|\uparrow\Uparrow\uparrow\Uparrow\rangle, \quad |u\bar{d}\rangle\frac{|\uparrow\Uparrow\downarrow\Downarrow\rangle + |\downarrow\Downarrow\uparrow\Uparrow\rangle}{\sqrt{2}}, \quad |u\bar{d}\rangle|\downarrow\Downarrow\downarrow\Downarrow\rangle; \quad m_\rho = 2m_q + V_S + A_M .$$

$$(10.20)$$

Since the spin-parity of spin-1 mesons, like ρ meson, is $J^P = 1^-$, we call them Vector (V) meson and the spin-parity of spin-0 mesons, like π meson is $J^P = 0^-$, we call them Pseudoscaler (PS) meson.

We can extract the strength of the transition amplitude A_M and the spin-independent strong potential V_S from the hadron masses

$$A_M = \frac{m_\rho - m_\pi}{4} \sim 160 \text{ MeV}, \quad V_S = \frac{3m_\rho + m_\pi}{4} - 2m_q \sim -80 \text{ MeV} . \quad (10.21)$$

Figure 10.6 shows the level structure of this $|u\bar{d}\rangle$ system.

The A_M value is very much different from the $A_H (\sim 1.5 \times 10^{-6} \text{eV})$ of the hydrogen atom in Chap. 3. However, this difference can be explained by the difference in the sizes, the coupling constants, and the constituent fermion masses between the meson and hydrogen atom. The ratio of A_M and A_H after correcting by these parameters is close to unity in spite of the large difference of the size and interactions, as shown below.

$$\frac{A_M}{A_H}\left(\frac{\alpha}{\alpha_S}\right)\left(\frac{a_M}{a_B}\right)^3\left(\frac{m_q^2}{m_e m_p}\right)\kappa_p \sim O[1] , \quad (10.22)$$

where $\kappa_p (\sim 2.8)$ is the factor of the proton anomalous magnetic moment. This indicates that the mechanism of the mass splitting between π^+ and ρ^+ is the same as that of the hyperfine-level splitting of the hydrogen atom.

10.4   Structure of ρ^0, ω and ϕ

10.4.1 Mixing Between $|u\bar{u}\rangle$ and $|d\bar{d}\rangle$

ρ^0, ω are spin-1 mesons which consist of $|u\bar{u}\rangle$, $|d\bar{d}\rangle$ states. From the isosymmetry of u and d-quarks, there are the same static and spin-dependent transitions as in π^+ and ρ^+ meson as shown in Fig. 10.7,

$$m_V \equiv 2m_q + V_S + A_M (= m_{\rho^+}) . \tag{10.23}$$

In addition to the processes shown in Fig. 10.7, there are annihilation process in ρ^0, and ω mesons as shown in Fig. 10.8. Due to the annihilation, flavor changing process $|u\bar{u}\rangle \leftrightarrow |d\bar{d}\rangle$ takes place in addition to the self-transition with equal amplitude (V_A). All the transition amplitudes between $|u\bar{u}\rangle$ and $|d\bar{d}\rangle$ states are summarized in Fig. 10.9. If we express the general wave function of $|u\bar{u}\rangle$ and $|d\bar{d}\rangle$ state as

$$|\psi_{UD}[t]\rangle = C_U[t] |u\bar{u}\rangle + C_D[t] |d\bar{d}\rangle , \tag{10.24}$$

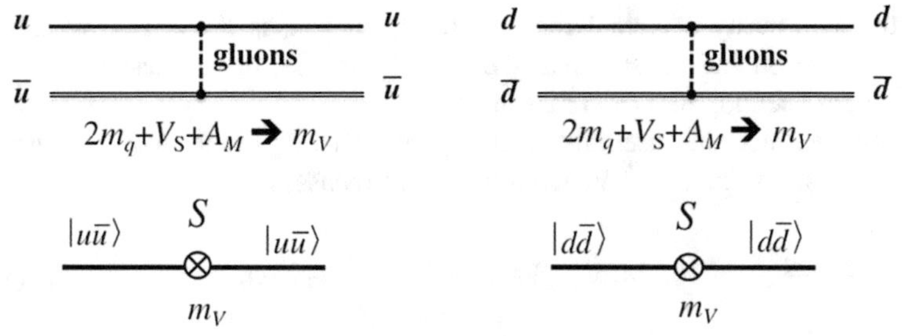

Fig. 10.7 $u\bar{u}$ and $d\bar{d}$ self-transitions. The line labeled "gluons" represents the effect of one or more gluon exchanges, not just one gluon. The quark mass, strong potential, dipole moment interaction are the same as the ρ^+ case. We define $m_V \equiv 2m_q + V_S + A_M$

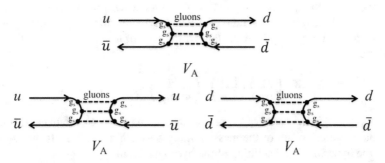

Fig. 10.8 $u\bar{u}$ and $d\bar{d}$ annihilation process. In order to conserve the color and the quantum number, $J^{PC} = 1^{--}$, at least three gluons are necessary

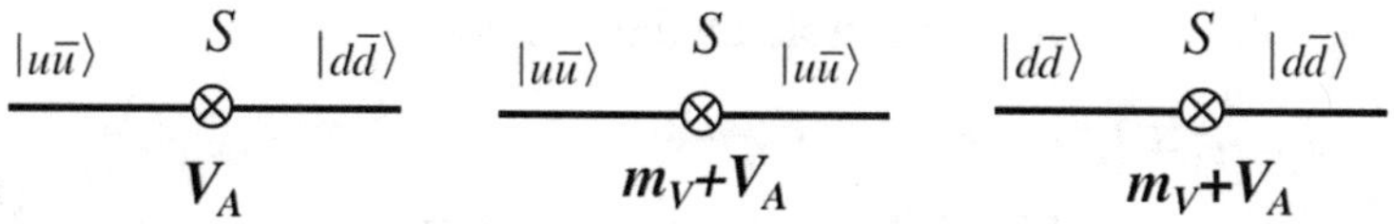

$$
\begin{array}{ccc}
\underset{\displaystyle V_A}{\underline{|u\bar{u}\rangle \overset{S}{\otimes} |d\bar{d}\rangle}} & \underset{\displaystyle m_V + V_A}{\underline{|u\bar{u}\rangle \overset{S}{\otimes} |u\bar{u}\rangle}} & \underset{\displaystyle m_V + V_A}{\underline{|d\bar{d}\rangle \overset{S}{\otimes} |d\bar{d}\rangle}}
\end{array}
$$

Fig. 10.9 $u\bar{u}$ and $d\bar{d}$ transition amplitudes

the transition equation becomes

$$
i\frac{d}{dt}\begin{pmatrix} C_U \\ C_D \end{pmatrix} = \begin{pmatrix} m_V + V_A & V_A \\ V_A & m_V + V_A \end{pmatrix}\begin{pmatrix} C_U \\ C_D \end{pmatrix} . \tag{10.25}
$$

10.4.1.1   The Mass Eigenstate
From Eq. (10.25), the mass eigenstates are

$$
\begin{cases} |\psi_{\mathrm{I}}\rangle = \frac{1}{\sqrt{2}}(|u\bar{u}\rangle - |d\bar{d}\rangle)e^{-im_V t} \Rightarrow \rho^0 \\ |\psi_{\mathrm{II}}\rangle = \frac{1}{\sqrt{2}}(|u\bar{u}\rangle + |d\bar{d}\rangle)e^{-i(m_V + 2V_A)t} \Rightarrow \omega . \end{cases} \tag{10.26}
$$

The state ψ_{I} has the same mass as ρ^+ as expected and we identify it as ρ^0 meson. Then the another mass eigenstate ψ_{II} as ω meson. From the mass relations,

$$
m_V = m_\rho \sim 775 \text{ MeV}, \qquad m_V + 2V_A = m_\omega \sim 783 \text{ MeV} . \tag{10.27}
$$

It is possible to obtain the annihilation amplitude V_A,

$$
V_A = \frac{m_\omega - m_\rho}{2} \sim 4 \text{ MeV} . \tag{10.28}
$$

Note that in the vector meson case, V_A is much smaller than the potential V_S and the quark mass m_q.

10.4.1.2 $|u\bar{u}\rangle \leftrightarrow |d\bar{d}\rangle$ Oscillation
In a e^+e^- experiment, u-$\bar{u}$ pair can be produced by the following interaction:

$$
e^+ + e^- \to \gamma^* \to u + \bar{u} . \tag{10.29}
$$

Now we assume that the center of mass energy is equivalent to the ρ^0 and ω masses.[2] As soon as the u-$\bar{u}$ pair is generated, the quarks start to interact with each other strongly with the processes shown in Figs. 10.7 and 10.8. The development of the wave function in time can be obtained by taking the initial condition

$$
|\psi_{UD}[0]\rangle = |u\bar{u}\rangle = \frac{1}{\sqrt{2}}(|\rho^0\rangle + |\omega\rangle) . \tag{10.30}
$$

---

[2]Their masses have broad width and the two-state overlap in energy.

In this case, the wave function at arbitrary time t becomes

$$\psi_{\text{UD}}[t] = \left|\rho^0\right\rangle e^{-im_\rho t} + |\omega\rangle e^{-im_\omega t}$$
$$= e^{-i(m_V + V_A)t}\left(\cos[V_A t]\left|u\bar{u}\right\rangle - i\sin[V_A t]\left|d\bar{d}\right\rangle\right). \tag{10.31}$$

The probability that the system is in $\left|d\bar{d}\right\rangle$ state at time t is

$$P_{\left|u\bar{u}\right\rangle \to \left|d\bar{d}\right\rangle}[t] = \left|\langle d\bar{d}|\psi_{\text{UD}}[t]\rangle\right|^2 = \sin^2[A_V t]. \tag{10.32}$$

The probability oscillates with oscillation period $\tau = \pi/A_V \sim 5 \times 10^{-24}$ s. It is so quick that we cannot measure the flavor of the quark before the flavor turns to another flavor. We will always observe $|u\bar{u}\rangle$ with 50% probability and $\left|d\bar{d}\right\rangle$ with 50% probability. This is the reason why there is no meson with pure $|u\bar{u}\rangle$ or $\left|d\bar{d}\right\rangle$ system.

10.4.2 Mixing Between $|s\bar{s}\rangle$ and $|u\bar{u}\rangle$, $|d\bar{d}\rangle$ Systems

Although there are no $|u\bar{u}\rangle$ nor $\left|d\bar{d}\right\rangle$ mesons, there is $|s\bar{s}\rangle$ meson called ϕ meson. Why is it? s-quark mass is $\sim$150 MeV heavier than u, d-quark masses. The transitions between $|u\bar{u}\rangle$, $\left|d\bar{d}\right\rangle$ system and $|s\bar{s}\rangle$ are shown in Fig. 10.10. Since the magnetic moment is proportional to $1/m_f$, the spin-dependent amplitude for $s\bar{s}$ system: A'_M is assumed to be smaller than A_M

$$A'_M = \left(\frac{m_q}{m_s}\right)^2 A_M \sim \frac{1}{2}A_M \sim 80 \text{ MeV}. \tag{10.33}$$

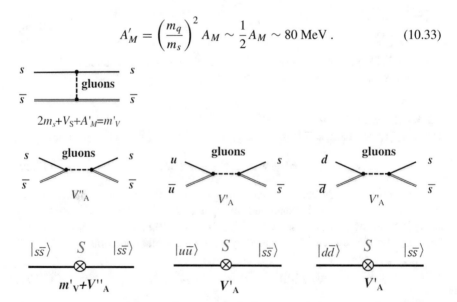

Fig. 10.10 Transition diagrams and amplitudes which include $|s\bar{s}\rangle$ system. We assume V_S, A_M, and V_A are the same for s-quarks related diagrams. The line labeled "gluons" represents the effect of one or more gluon exchanges as shown in Fig. 10.8

As for the annihilation amplitudes, V'_A and V''_A are as small as $V_A \sim 4$ MeV and much smaller than the mass difference between s and u, d quarks.

From Fig. 10.10, it is possible to show that the transition between $(|u\bar{u}\rangle - |d\bar{d}\rangle)/\sqrt{2}$ system and $|s\bar{s}\rangle$ system does not happen

$$\langle \frac{1}{\sqrt{2}}(u\bar{u} - d\bar{d})|H_S|s\bar{s}\rangle = \frac{1}{\sqrt{2}}(\langle u\bar{u}|H_S|s\bar{s}\rangle - \langle d\bar{d}|H_S|s\bar{s}\rangle) = \frac{1}{\sqrt{2}}(V'_A - V'_A) = 0 .$$

(10.34)

This is due to the fact that d and u quarks are symmetric in strong interactions. The transition between $|u\bar{u}\rangle$, $|d\bar{d}\rangle$ system and $|s\bar{s}\rangle$ system happens as transition between $(|u\bar{u}\rangle + |d\bar{d}\rangle)/\sqrt{2}$ and $|s\bar{s}\rangle$. The transition amplitude is

$$\langle \frac{1}{\sqrt{2}}(u\bar{u} + d\bar{d})|H_S|s\bar{s}\rangle = \frac{1}{\sqrt{2}}(\langle u\bar{u}|H_S|s\bar{s}\rangle + \langle d\bar{d}|H_S|s\bar{s}\rangle) = \frac{1}{\sqrt{2}}(V'_A + V'_A) = \sqrt{2}V'_A ,$$

(10.35)

where H_S is the strong interaction Hamiltonian. These properties are true for any three-body mixing systems in which two of the components are symmetric.

Figure 10.11 shows the transition amplitudes between $|\pm\rangle = (|u\bar{u}\rangle \pm |d\bar{d}\rangle)/\sqrt{2}$ systems and $|s\bar{s}\rangle$.

Now we define the wave function as

$$|\psi_{UDS}[t]\rangle = C_U[t]\,|u\bar{u}\rangle + C_D[t]\,|d\bar{d}\rangle + C_S[t]\,|s\bar{s}\rangle .$$

(10.36)

This wave function can be rewritten as

$$|\psi_{UDS}\rangle = \frac{C_U - C_D}{\sqrt{2}}\frac{|u\bar{u}\rangle - |d\bar{d}\rangle}{\sqrt{2}} + \frac{C_U + C_D}{\sqrt{2}}\frac{|u\bar{u}\rangle + |d\bar{d}\rangle}{\sqrt{2}} + C_S\,|s\bar{s}\rangle$$

$$\equiv C_-[t]\,|-\rangle + C_+[t]\,|+\rangle + C_S[t]\,|s\bar{s}\rangle .$$

(10.37)

The transitions between $|+\rangle$ and $|-\rangle$ states are shown in Fig. 10.12. Since $|-\rangle$ is decoupled from $|s\bar{s}\rangle$ and $|+\rangle$, we can adopt a result of previous section. From Eq. (10.26), $|-\rangle$ is mass eigenstate and corresponds to ρ^0 meson,

$$|\rho^0[t]\rangle = e^{-im_V t}\,|-\rangle .$$

(10.38)

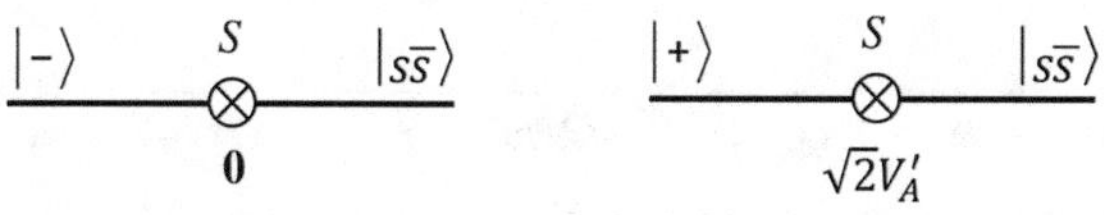

Fig. 10.11 Transitions between $(|u\bar{u}\rangle \pm |d\bar{d}\rangle)/\sqrt{2}$ states and $|s\bar{s}\rangle$. The transition amplitude between $(|u\bar{u}\rangle - |d\bar{d}\rangle)/\sqrt{2}$ and $|s\bar{s}\rangle$ is 0

Fig. 10.12 Self-transitions of $|\pm\rangle$. Taken from Eq. (10.26)

Fig. 10.13 Mixing triangle for the transition Eq. (10.39) and its mass eigenstates (10.40)

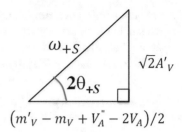

$$(m'_V - m_V + V''_A - 2V_A)/2$$

The transition equation of $|+\rangle$ and $|s\bar{s}\rangle$ is

$$i\frac{d}{dt}\begin{pmatrix} C_S \\ C_+ \end{pmatrix} = \begin{pmatrix} m'_V + V''_A & \sqrt{2}V'_A \\ \sqrt{2}V'_A & m_V + 2V_A \end{pmatrix}\begin{pmatrix} C_S \\ C_+ \end{pmatrix}, \tag{10.39}$$

where $m'_V = 2m_s + V_S + A'_M$ and $m_V = 2m_q + V_S + A_M$. The wave functions of the mass eigenstate are

$$\begin{cases} |\psi_\omega\rangle = (\cos\theta_{+S}|+\rangle + \sin\theta_{+S}|s\bar{s}\rangle)e^{-i(\bar{\mu}-\omega)t} \\ |\psi_\phi\rangle = (-\sin\theta_{+S}|+\rangle + \cos\theta_{+S}|s\bar{s}\rangle)e^{-i(\bar{\mu}+\omega)t} \end{cases} \tag{10.40}$$

The corresponding mixing triangle is shown in Fig. 10.13. The parameters in Eq. (10.40) are expressed by the transition amplitudes as

$$\tan 2\theta_{+S} = \frac{\sqrt{2}V'_A}{m_s - m_q + (A'_M - A_M)/2 - V_A + V''_A/2}$$

$$\omega_{+S} = \sqrt{\left(m_s - m_q + (A'_M - A_M)/2 - V_A + V''_A/2\right)^2 + 2V'^2_A} \tag{10.41}$$

$$\bar{\mu}_{+S} = m_s + m_q + V_S + V_A + V''_A/2 + (A_M + A'_M)/2.$$

For ρ^0, ω, ϕ case, there are the relations:

$$V'_A \sim V''_A \sim V_A(\sim 4\,\text{MeV}) \ll m_s - m_q(\sim 150\,\text{MeV})$$
$$A_M \sim 2A'_M \sim 160\,\text{MeV}, \tag{10.42}$$

and the parameters in (10.41) can be approximated as

$$\theta_{+S} \sim \frac{V'_A}{\sqrt{2}(m_s - m_q - A_M/4)} \sim \frac{V_A}{\sqrt{2}(m_s - m_q - A_M/4)} \sim 0.026,$$

$$\omega_{+S} \sim m_s - m_q - A_M/4 \sim 110\,\text{MeV}, \tag{10.43}$$

$$\bar{\mu}_{+S} \sim m_s + m_q + V_S + \frac{3}{4}A_M \sim 890\,\text{MeV}.$$

The probability to find $|s\bar{s}\rangle$ in ω meson and to find $|u\bar{u}\rangle$ or $|d\bar{d}\rangle$ in ϕ meson is very small as

$$P = \sin^2\theta_{+S} \sim 0.0007. \tag{10.44}$$

Usually, we ignore this component and write their quark structures as

$$|\omega\rangle = \frac{|u\bar{u}\rangle + |d\bar{d}\rangle}{\sqrt{2}}, \quad |\phi\rangle = |s\bar{s}\rangle \ . \tag{10.45}$$

The reason that the pure $|s\bar{s}\rangle$ state can be a meson (=energy eigenstate), unlike $|u\bar{u}\rangle$ or $|d\bar{d}\rangle$ cases, is that the transition amplitude between $q\bar{q}$ and $s\bar{s}$ system is much smaller than their mass difference.

The mass is predicted from this simple model as

$$\begin{cases} m_\omega = \bar{\mu} - \omega_{+S} \sim 780 \text{ MeV} \\ m_\phi = \bar{\mu} + \omega_{+S} \sim 1000 \text{ MeV} \ . \end{cases} \tag{10.46}$$

These are similar to the observed values: $m_\omega = 782.7$ MeV and $m_\phi = 1019.5$ MeV.

10.4.3 Experimental Confirmation of the Vector Meson Structure

The quark structure of the vector mesons derived above can be tested experimentally. The vector mesons decay to e^+e^- pair through photon as shown in Fig. 10.14. Since the quark structures of the vector mesons are Eqs. (10.38) and (10.45), the expected

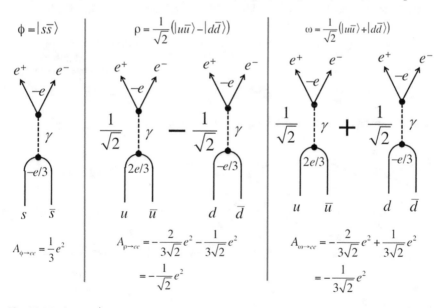

Fig. 10.14 $\psi_V \to e^+e^-$ decay diagrams

relative decay widths are, ignoring the mass difference,

$$\Gamma_{\rho \to ee} : \Gamma_{\omega \to ee} : \Gamma_{\phi \to ee}$$

$$\sim \left| \frac{1}{\sqrt{2}} \left(-\frac{2}{3} e^2 - \frac{1}{3} e^2 \right) \right|^2 : \left| \frac{1}{\sqrt{2}} \left(-\frac{2}{3} e^2 + \frac{1}{3} e^2 \right) \right|^2 : \left| \frac{1}{3} e^2 \right|^2 = 9 : 1 : 2 .$$

$$(10.47)$$

On the other hand, if there are no quark mixings, the expected decay widths are

$$\Gamma_{|u\bar{u}\rangle \to ee} : \Gamma_{|d\bar{d}\rangle \to ee} : \Gamma_{|s\bar{s}\rangle \to ee} \sim \left| \frac{2}{3} e^2 \right|^2 : \left| -\frac{1}{3} e^2 \right|^2 : \left| -\frac{1}{3} e^2 \right|^2 = 4 : 1 : 1 .$$

$$(10.48)$$

The observations $\Gamma_{\rho \to ee} : \Gamma_{\omega \to ee} : \Gamma_{\phi \to ee} \sim 11.3 : 1 : 2.3$ support the hypothesis (10.47) and the discussions on the mixings and oscillations are confirmed right.

10.5   Structure of π^0, η and η'

Like ρ^0, ω and ϕ mesons, π^0, η and η' mesons consist of $|u\bar{u}\rangle$, $|d\bar{d}\rangle$ and $|s\bar{s}\rangle$, too. Since the spin of π^0, η and η' are S=0, the spin-dependent potential is $-3A_M$ for $|u\bar{u}\rangle$ and $|d\bar{d}\rangle$ and $-3A'_M$ for $|s\bar{s}\rangle$ system. The strong interaction potentials are shown in Fig. 10.15. There are also annihilation diagrams as shown in Fig. 10.16. The transition amplitudes between $|u\bar{u}\rangle$, $|d\bar{d}\rangle$ and $|s\bar{s}\rangle$ states are summarized in Fig. 10.17.

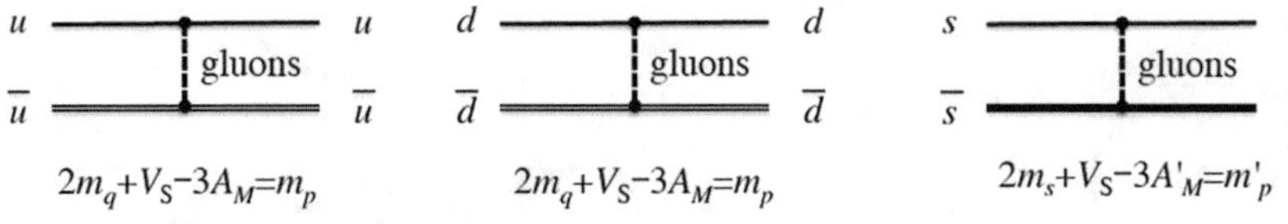

$$
\begin{array}{ccc}
u \,\rule[0.5ex]{2em}{0.4pt}\!\!\bullet\!\!\rule[0.5ex]{2em}{0.4pt}\, u & \quad d \,\rule[0.5ex]{2em}{0.4pt}\!\!\bullet\!\!\rule[0.5ex]{2em}{0.4pt}\, d & \quad s \,\rule[0.5ex]{2em}{0.4pt}\!\!\bullet\!\!\rule[0.5ex]{2em}{0.4pt}\, s \\
\text{gluons} & \text{gluons} & \text{gluons} \\
\bar{u} \,\rule[0.5ex]{2em}{0.4pt}\!\!\bullet\!\!\rule[0.5ex]{2em}{0.4pt}\, \bar{u} & \bar{d} \,\rule[0.5ex]{2em}{0.4pt}\!\!\bullet\!\!\rule[0.5ex]{2em}{0.4pt}\, \bar{d} & \bar{s} \,\rule[0.5ex]{2em}{0.4pt}\!\!\bullet\!\!\rule[0.5ex]{2em}{0.4pt}\, \bar{s} \\
2m_q + V_S - 3A_M = m_p & 2m_q + V_S - 3A_M = m_p & 2m_s + V_S - 3A'_M = m'_p
\end{array}
$$

Fig. 10.15 Strong potential for $u\bar{u}$, $d\bar{d}$ and $s\bar{s}$ pseudoscalar mesons

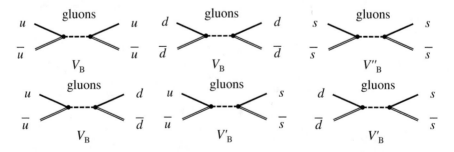

Fig. 10.16 Annihilation process of $u\bar{u}$, $d\bar{d}$ and $s\bar{s}$

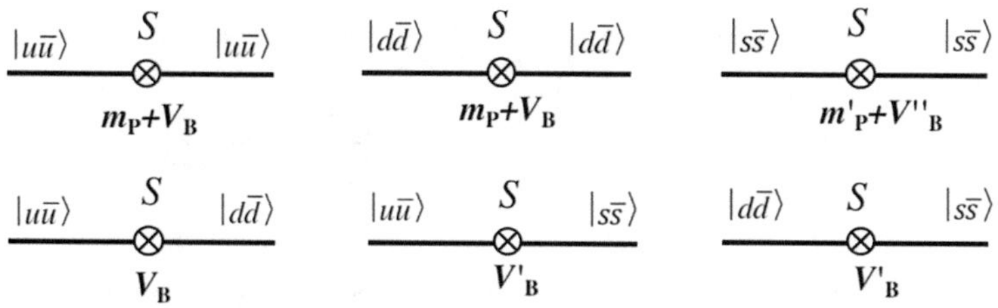

$$\underline{|u\bar{u}\rangle \quad \overset{S}{\underset{\bigotimes}{}} \quad |u\bar{u}\rangle} \qquad \underline{|d\bar{d}\rangle \quad \overset{S}{\underset{\bigotimes}{}} \quad |d\bar{d}\rangle} \qquad \underline{|s\bar{s}\rangle \quad \overset{S}{\underset{\bigotimes}{}} \quad |s\bar{s}\rangle}$$
$$m_P + V_B \qquad\qquad m_P + V_B \qquad\qquad m'_P + V''_B$$

$$\underline{|u\bar{u}\rangle \quad \overset{S}{\underset{\bigotimes}{}} \quad |d\bar{d}\rangle} \qquad \underline{|u\bar{u}\rangle \quad \overset{S}{\underset{\bigotimes}{}} \quad |s\bar{s}\rangle} \qquad \underline{|d\bar{d}\rangle \quad \overset{S}{\underset{\bigotimes}{}} \quad |s\bar{s}\rangle}$$
$$V_B \qquad\qquad\qquad V'_B \qquad\qquad\qquad V'_B$$

Fig. 10.17 Amplitudes of transitions between $u\bar{u}$, $d\bar{d}$ and $s\bar{s}$

So far, qualitative discussions are the same as the ρ^0, ω, ϕ cases. From the analogy to Eq. (10.38), the π^0 wave function is,

$$\psi_{\pi^0} = e^{-im_P t} \frac{|u\bar{u}\rangle - |d\bar{d}\rangle}{\sqrt{2}} . \tag{10.49}$$

The small mass difference between π^0 and π^+ can be attributed to the difference of the quark masses and charges.

From here, there is significant difference between the pseudoscalar and vector mesons. If the quark structure of η meson was $(|u\bar{u}\rangle + |d\bar{d}\rangle)/\sqrt{2}$, like ω meson, the annihilation amplitude would be, from the analogy of Eq. (10.28),

$$V_B = \frac{m_\eta - m_{\pi^0}}{2} = \frac{548 - 135}{2} \sim 200 \text{ MeV} . \tag{10.50}$$

This is larger than the mass difference between s and d quarks and V'_B may not be small enough to be ignored as vector meson case ($V_A \sim 4$ MeV). If V'_B is not so small, we need to take into account the mixing between $|s\bar{s}\rangle$ state and $|+\rangle$ state,

$$\begin{cases} |\eta\rangle = \cos\theta_P \dfrac{|u\bar{u}\rangle + |d\bar{d}\rangle}{\sqrt{2}} + \sin\theta_P \, |s\bar{s}\rangle \\ |\eta'\rangle = -\sin\theta_P \dfrac{|u\bar{u}\rangle + |d\bar{d}\rangle}{\sqrt{2}} + \cos\theta_P \, |s\bar{s}\rangle . \end{cases} \tag{10.51}$$

Experimentally, θ_P is measured as $36.5 \pm 0.8°$[2].

The reason why quark annihilation in PS meson is much stronger than that in vector meson can be qualitatively understood as follows. In the PS meson, at least two gluons are necessary for annihilation as shown in Fig. 10.18, while in vector meson, at least three gluons are necessary as shown in Fig. 10.8 to conserve color and the J^{PC} quantum number. Therefore, the annihilation amplitude for PS meson is proportional to α_S^2 and for V meson, α_S^3. In addition, the average energy per gluon in π^0 is 70 MeV, while in ρ^0 meson, it is 250 MeV and the running strong coupling constant $\alpha_S[Q]$ is expected to be smaller for vector meson than in PS meson as indicated in Fig. 10.4.

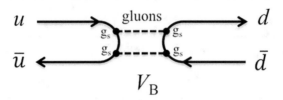

Fig. 10.18 Annihilation process of $u\bar{u}$, $d\bar{d}$ in PS meson. At least two gluons have to be exchanged to conserve the color and the quantum number, $J^{PC} = 0^{-+}$

10.6   Color Structure of Meson

Quark has color quantum number and antiquark has anti-color;

$$q_R, \ q_G, \ q_B, \ \ \bar{q}_R, \ \bar{q}_G, \ \bar{q}_B \ . \tag{10.52}$$

On the other hand, a gluon has both color and anticolor,

$$\begin{aligned} &G_{R\bar{G}}, \ G_{R\bar{B}}, \ G_{G\bar{R}}, \ G_{G\bar{B}}, \ G_{B\bar{R}}, \ G_{B\bar{G}}, \\ &G_- \left(= \frac{G_{R\bar{R}} - G_{G\bar{G}}}{\sqrt{2}} \right), \ \ G_+ \left(= \frac{G_{R\bar{R}} + G_{G\bar{G}} - 2G_{B\bar{B}}}{\sqrt{6}} \right) \ . \end{aligned} \tag{10.53}$$

The amplitude of a color pair changing process is $A(G\bar{G} \to R\bar{R}) = -g_S^2$ as shown in Fig. 10.19a. There are two kinds of process for color conserving process (Fig. 10.19b) and the total amplitude is $A(G\bar{G} \to G\bar{G}) = -2g_S^2/3$. If we write the wave function of this meson as

$$\psi_{q\bar{q}'}[t] = \left| q\bar{q}' \right\rangle (C_{R\bar{R}}[t] \left| R\bar{R} \right\rangle + C_{G\bar{G}}[t] \left| G\bar{G} \right\rangle + C_{B\bar{B}}[t] \left| B\bar{B} \right\rangle) \ . \tag{10.54}$$

The transition equation is

$$i\frac{d}{dt} \begin{pmatrix} C_{R\bar{R}} \\ C_{G\bar{G}} \\ C_{B\bar{B}} \end{pmatrix} = -g_S^2 \begin{pmatrix} 2/3 & 1 & 1 \\ 1 & 2/3 & 1 \\ 1 & 1 & 2/3 \end{pmatrix} \begin{pmatrix} C_{R\bar{R}} \\ C_{G\bar{G}} \\ C_{B\bar{B}} \end{pmatrix} \ . \tag{10.55}$$

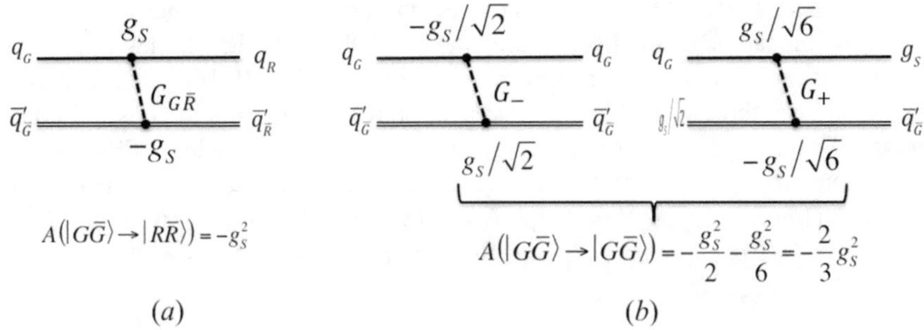

$q_G \rule{2cm}{0.4pt}^{\ g_S}\ q_R \qquad q_G \rule{2cm}{0.4pt}^{\ -g_S/\sqrt{2}}\ q_G \qquad q_G \rule{2cm}{0.4pt}^{\ g_S/\sqrt{6}}\ g_S$

$\bar{q}'_{\bar{G}} \cdots\cdots \bar{q}_{\bar{R}} \qquad \bar{q}'_{\bar{G}} \cdots\cdots \bar{q}_{\bar{G}} \qquad \bar{q}'_{\bar{G}} \cdots\cdots \bar{q}'_{\bar{G}}$

with $G_{G\bar{R}}$, $-g_S$; G_-, $g_S/\sqrt{2}$; G_+, $-g_S/\sqrt{6}$

$A(|G\bar{G}\rangle \to |R\bar{R}\rangle) = -g_S^2 \qquad\qquad A(|G\bar{G}\rangle \to |G\bar{G}\rangle) = -\frac{g_S^2}{2} - \frac{g_S^2}{6} = -\frac{2}{3}g_S^2$

$\qquad\qquad (a) \qquad\qquad\qquad\qquad\qquad\qquad\qquad (b)$

Fig. 10.19 Color transition. **a** Color changing process. **b** Color unchanging process

From this, the mass eigenstates are

$$
\begin{cases}
|\psi_1[t]\rangle &= |q\overline{q'}\rangle \dfrac{|R\overline{R}\rangle+|G\overline{G}\rangle+|B\overline{B}\rangle}{\sqrt{3}} \exp\left[i\frac{8}{3}g_S^2 t\right] \\[2mm]
|\psi_2[t]\rangle &= |q\overline{q'}\rangle \dfrac{|R\overline{R}\rangle-|G\overline{G}\rangle}{\sqrt{2}} \exp\left[-i\frac{1}{3}g_S^2 t\right] \\[2mm]
|\psi_3[t]\rangle &= |q\overline{q'}\rangle \dfrac{|R\overline{R}\rangle-|B\overline{B}\rangle}{\sqrt{2}} \exp\left[-i\frac{1}{3}g_S^2 t\right] .
\end{cases}
\tag{10.56}
$$

The mass (potential energy) of ψ_1 is negative and the color force between quarks of this state is attractive. The energies of ψ_1 and ψ_2 are positive and the color force of these states is repulsive. Therefore, the color structure of the meson is ψ_1.

The followings are wave functions with the color components for some mesons:

$$
|\pi^+\rangle = |u\overline{d}\rangle \frac{|\Uparrow\Downarrow\rangle - |\Downarrow\Uparrow\rangle}{\sqrt{2}} \frac{|R\overline{R}\rangle + |G\overline{G}\rangle + |B\overline{B}\rangle}{\sqrt{3}} ,
$$

$$
|\pi^0\rangle = \frac{|u\overline{u}\rangle - |d\overline{d}\rangle}{\sqrt{2}} \frac{|\Uparrow\Downarrow\rangle - |\Downarrow\Uparrow\rangle}{\sqrt{2}} \frac{|R\overline{R}\rangle + |G\overline{G}\rangle + |B\overline{B}\rangle}{\sqrt{3}} ,
\tag{10.57}
$$

$$
|\rho^+\rangle = |u\overline{d}\rangle \frac{|\Uparrow\Downarrow\rangle + |\Downarrow\Uparrow\rangle}{\sqrt{2}} \frac{|R\overline{R}\rangle + |G\overline{G}\rangle + |B\overline{B}\rangle}{\sqrt{3}} .
$$

References

1. Particle Data Group. Prog. Theor. Exp. Phys. **2020**, 083C01 (2020)
2. Balitsky, J.V., et al.: (2015). arXiv:1505.07750

Quark Structure of Baryons

11

11.1 Introduction

Baryons consist of three quarks, such as $|p\rangle = |uud\rangle$, $|\Lambda\rangle = |uds\rangle$, $|\Omega^-\rangle = |sss\rangle$. Table 11.1 shows quark combinations and corresponding baryons and their masses.

There are two types of baryon family. One is spin-1/2 baryons and the other is spin-3/2 baryons. In Table 11.1, we notice that for the spin-1/2 baryon family, $|uuu\rangle$, $|ddd\rangle$, $|sss\rangle$ baryons do not exist and there are two $|uds\rangle$ baryons. These properties can be understood from the mixing structure of the quark system in baryons.

11.2 Totally Antisymmetric State

When thinking of the quark structure of baryons, it is necessary to consider the Pauli exchange principle for fermions. It requires that if any pair of fermions are exchanged, the wave function has to change its sign,[1]

$$\psi[q_1, q_2] = -\psi[q_2, q_1] . \tag{11.1}$$

For baryon case, since it consists of three quarks, the antisymmetric principle requires

$$\psi_B[q_1, q_2, q_3] = -\psi_B[q_2, q_1, q_3] = \psi_B[q_2, q_3, q_1]$$
$$= -\psi_B[q_3, q_2, q_1] = \psi_B[q_3, q_1, q_2] = -\psi_B[q_1, q_3, q_2] . \tag{11.2}$$

This property is called *totally antisymmetric*. It plays very important roles to determine the baryon structure.

---

[1] This is also true for the meson cases. For example, strictly speaking, the quark flavor-spin structure of the π^+ meson is $|\pi^+\rangle = \frac{|u\bar{d}\rangle + |\bar{d}u\rangle}{\sqrt{2}} \frac{|\uparrow\Downarrow\rangle - |\Downarrow\uparrow\rangle}{\sqrt{2}}$.

© Springer Nature Switzerland AG 2021
F. Suekane, *Quantum Oscillations*, Lecture Notes in Physics 985,
https://doi.org/10.1007/978-3-030-70527-5_11

Table 11.1 Baryons consisting of u, d, s-quarks. "×" means no baryon exists with that quark combination and spin. $n_u + n_d + n_s = 3$

$S = 1/2$	$n_d = 0$	$n_d = 1$	$n_d = 2$	$n_d = 3$
$n_s = 0$	×	$p(938.3)$	$n(939.6)$	×
$n_s = 1$	$\Sigma^+(1189.4)$	$\Sigma^0(1192.6)$ $\Lambda(1115.7)$	$\Sigma^-(1197.4)$	
$n_s = 2$	$\Xi^0(1314.9)$	$\Xi^-(1321.7)$		
$n_s = 3$	×			

$S = 3/2$	$n_d = 0$	$n_d = 1$	$n_d = 2$	$n_d = 3$
$n_s = 0$	$\Delta^{++}(1232)$	Δ^+	Δ^0	Δ^-
$n_s = 1$	$\Sigma^{+*}(1383)$	$\Sigma^{0*}(1384)$	$\Sigma^{-*}(1387)$	
$n_s = 2$	$\Xi^{0*}(1532)$	$\Xi^{-*}(1535)$		
$n_s = 3$	$\Omega^-(1673)$			

11.3 Δ^{++} Baryon

Let us start with a baryon with simplest structure. Δ^{++} is spin-3/2, $|uuu\rangle$ baryon whose base state vector can be written as

$$|\Delta^{++}\rangle = |uuu\rangle \, |\Uparrow\Uparrow\Uparrow\rangle \, |c_1c_2c_3\rangle \,, \tag{11.3}$$

where c_i indicate the color. Since the flavor-spin part is symmetric for exchange of any two quarks, the color part of the wave function has to be totally antisymmetric. For example, $|c_2c_1c_3\rangle = -|c_1c_2c_3\rangle$. To satisfy this, c_1 and c_2 cannot be identical. Which means the color state includes all R, G, B color.

In baryons, the quark color changes by exchanging the gluons. The gluon itself carries color and anti-color as in Eq. (10.53). Figure 11.1a shows the process that the color state $|RG\rangle$ changes to $|GR\rangle$ by exchanging the gluon $G_{R\bar{G}}$. Therefore, we

Fig. 11.1 **a** $|RG\rangle \leftrightarrow |GR\rangle$ cross-transition. The amplitude is g_S^2. **b** $|RG\rangle \leftrightarrow |RG\rangle$ self-transition. The amplitude is $-g_S^2/3$

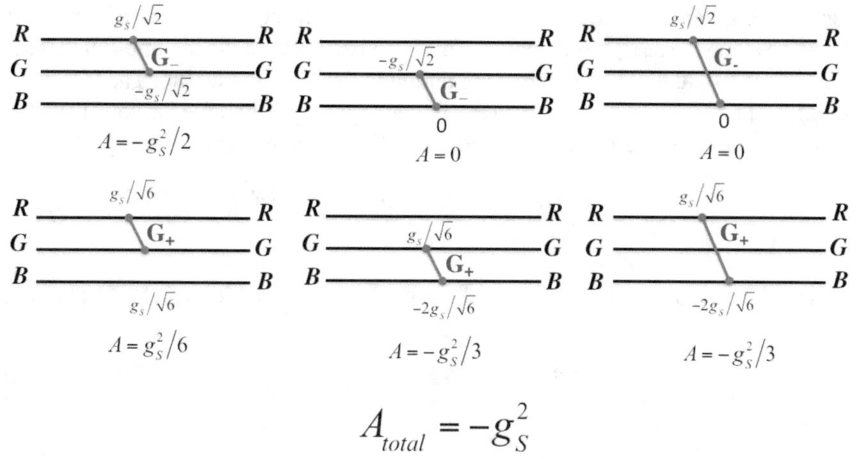

Fig. 11.2 Self-transition amplitude for three quark case, $-g_S^2$

can assume the color state is a superposition of the different color-order states as

$$|\psi[t]\rangle = |q_1q_2q_3\rangle(C_{RGB}[t]\,|RGB\rangle + C_{RBG}[t]\,|RBG\rangle + C_{BRG}[t]\,|BRG\rangle$$
$$+ C_{BGR}[t]\,|BGR\rangle + C_{GRB}[t]\,|GRB\rangle + C_{GBR}[t]\,|GBR\rangle). \tag{11.4}$$

Figure 11.1b shows the self-transition amplitude between two quarks, in which color neutral gluons are exchanged. For the three quark case, there are several self-transitions as shown in Fig. 11.2 and the total self-transition amplitude is $-g_S^2$. From Fig. 11.1a, the cross-transition amplitude is g_S^2. Finally, the color transition equation for three quarks is

$$i\frac{d}{dt}\begin{pmatrix} C_{RGB} \\ C_{RBG} \\ C_{BRG} \\ C_{BGR} \\ C_{GBR} \\ C_{GRB} \end{pmatrix} = g_S^2 \begin{pmatrix} -1 & 1 & 0 & 1 & 0 & 1 \\ 1 & -1 & 1 & 0 & 1 & 0 \\ 0 & 1 & -1 & 1 & 0 & 1 \\ 1 & 0 & 1 & -1 & 1 & 0 \\ 0 & 1 & 0 & 1 & -1 & 1 \\ 1 & 0 & 1 & 0 & 1 & -1 \end{pmatrix} \begin{pmatrix} C_{RGB} \\ C_{RBG} \\ C_{BRG} \\ C_{BGR} \\ C_{GBR} \\ C_{GRB} \end{pmatrix}. \tag{11.5}$$

There are six energy eigenstates. After some calculations, the wave function of the energy eigenstates are

$$\begin{cases} |\psi_A[t]\rangle = \frac{|RGB\rangle - |RBG\rangle + |BRG\rangle - |BGR\rangle + |GBR\rangle - |GRB\rangle}{\sqrt{6}} \exp[+4ig_s^2t] \\ |\psi_S[t]\rangle = \frac{|RGB\rangle + |RBG\rangle + |BRG\rangle + |BGR\rangle + |GBR\rangle + |GRB\rangle}{\sqrt{6}} \exp[-2ig_s^2t] \\ |\psi_3[t]\rangle = \frac{|GRB\rangle - |RBG\rangle}{\sqrt{2}} \exp[+ig_s^2t] \\ \vdots \end{cases} \tag{11.6}$$

The only totally antisymmetric state is ψ_A. The color force between quarks of ψ_A is attractive since the energy is negative. This is consistent with the assumption that the baryon is the state that three quarks are bound in the strong potential. The transition equation (11.5) is independent from the flavor structure and all the baryons have the same color structure of ψ_A. So, we name the color structure of ψ_A simply as

$$|\mathcal{C}_A\rangle \equiv \frac{|RGB\rangle - |RBG\rangle + |BRG\rangle - |BGR\rangle + |GBR\rangle - |GRB\rangle}{\sqrt{6}} \tag{11.7}$$

and use it hereafter for convenience. Finally, the wave function of Δ^{++} baryon is expressed as

$$\left|\Delta^{++}\right\rangle = |uuu\rangle \left|\Uparrow\Uparrow\Uparrow\right\rangle |\mathcal{C}_A\rangle . \tag{11.8}$$

11.3.1 Δ^+ Baryon

The quark contents of Δ^+ is (uud). In this case, the transition between u and d-quarks has to be taken into account. The transition between $|ud\rangle$ and $|du\rangle$ takes place by π^+ exchange as shown in Fig. 11.3a. π^+ meson consists of u and $\bar{d}$ quarks and in Fig. 11.3a, u-quark in the baryon emits π^+ and changes to d quark. The original d-quark absorbs the π^+ and changes to u quark and $|ud\rangle$ state changes to $|du\rangle$ state. Since the spin and net color of π^+ is 0, the spin and color of the original quarks do not change. There is also self-transition amplitude; $|ud\rangle \rightarrow |ud\rangle$ by π^0 exchange as shown in Fig. 11.3b. In the intermediate π^0, u and $\bar{u}$ immediately annihilate and pair-create $d\bar{d}$ state and the $\bar{d}$ and the original d annihilate leaving a d-quark. There

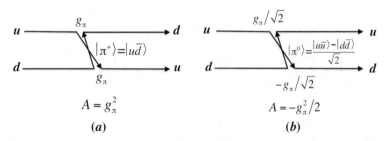

Fig. 11.3 Diagram of π meson exchange process between quarks. The time flows from left to right. **a** $|ud\rangle \rightarrow |du\rangle$ transition by π^+ exchange. The arrow on quark line indicates the propagation direction of the quark. The intermediate d-quark propagating backward in time is equivalent to the $\bar{d}$-quark propagating forward in time. g_π is quark-pion coupling constant. **b** $|ud\rangle \rightarrow |ud\rangle$ self-transition by π^0 exchange. π^0 is emitted from u-quark as $|u\bar{u}\rangle$. The u and $\bar{u}$ quarks immediately annihilate with each other by the strong interaction and changes to $|d\bar{d}\rangle$ ($|u\bar{u}\rangle \Leftrightarrow |d\bar{d}\rangle$ oscillation.) The original d-quark absorbs the π^0 staying as d-quark. Since the amplitude of $|u\bar{u}\rangle$ and $|d\bar{d}\rangle$ component in π^0 are $\pm 1/\sqrt{2}$, the coupling to $u\bar{u}$ and $d\bar{d}$ become $\pm g_\pi/\sqrt{2}$, respectively and the net $|u\bar{u}\rangle \rightarrow |u\bar{u}\rangle$ transition amplitude is $-g_\pi^2/2$

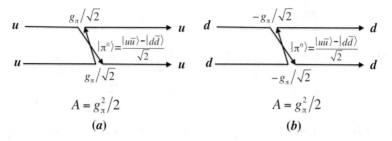

Fig. 11.4 a $|uu\rangle \to |uu\rangle$, **b** $|dd\rangle \to |dd\rangle$ transitions. The sign of the $d - \pi^0$ coupling is negative but the total amplitude is the same as the $|uu\rangle$ self-transition case

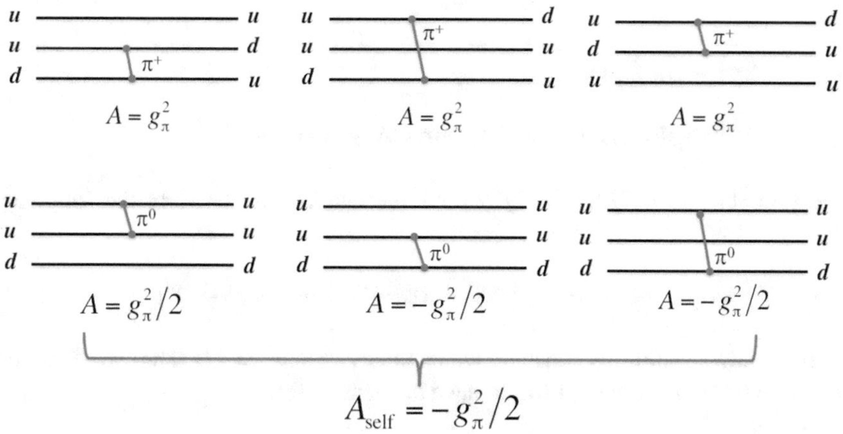

Fig. 11.5 $|uud\rangle$ transitions and their amplitudes. The amplitude for $|uud\rangle$ system is the same as that of $|udu\rangle$ and $|ddu\rangle$ self-amplitudes

is not change of quark flavor, spin, color but the energy level shifts due to the π^0 exchange effect. There are also $|uu\rangle \to |uu\rangle$ and $|dd\rangle \to |dd\rangle$ self-transitions as shown in Fig. 11.4.

For (uud) quark system, the flavor changing diagrams are summarized in Fig. 11.5. By writing the wave function as

$$|\psi[t]\rangle = C_{uud}[t]\,|uud\rangle + C_{udu}[t]\,|udu\rangle + C_{duu}[t]\,|duu\rangle , \qquad (11.9)$$

the transition equation is,

$$i\frac{d}{dt}\begin{pmatrix} C_{uud} \\ C_{udu} \\ C_{duu} \end{pmatrix} = g_\pi^2 \begin{pmatrix} -1/2 & 1 & 1 \\ 1 & -1/2 & 1 \\ 1 & 1 & -1/2 \end{pmatrix} \begin{pmatrix} C_{uud} \\ C_{udu} \\ C_{duu} \end{pmatrix} . \qquad (11.10)$$

The mass eigenstates are

$$|\psi_I\rangle = \frac{|uud\rangle + |udu\rangle + |duu\rangle}{\sqrt{3}} \exp\left[-i\frac{3}{2}g_\pi^2 t\right] ,$$

$$|\psi_{II}\rangle = \frac{|uud\rangle - |udu\rangle}{\sqrt{2}} \exp\left[i\frac{3}{2}g_\pi^2 t\right] , \quad |\psi_{III}\rangle = \frac{|uud\rangle - |duu\rangle}{\sqrt{2}} \exp\left[i\frac{3}{2}g_\pi^2 t\right] .$$

$$(11.11)$$

ψ_I is totally symmetric and the wave function of Δ^+ baryon is determined as follows.

$$|\Delta^+\rangle = \frac{|uud\rangle + |udu\rangle + |duu\rangle}{\sqrt{3}} |\uparrow\uparrow\uparrow\rangle |C_A\rangle \qquad (11.12)$$

11.4   Spin-1/2 Baryon

11.4.1 Why Spin-1/2 (uuu) Baryon Does Not Exist?

The spin state $|S = 1/2, s_z = 1/2\rangle$ for three quarks has generally the following structure,

$$|S, S_z\rangle = |1/2, +1/2\rangle = a(|\uparrow\uparrow\downarrow\rangle - |\downarrow\uparrow\uparrow\rangle) + b(|\uparrow\downarrow\uparrow\rangle - |\downarrow\uparrow\uparrow\rangle) , \quad (11.13)$$

where a and b are arbitrary constants which satisfy $a^2 + b^2 = 1/2$. The wave function of the spin-1/2 $|uuu\rangle$ baryon would have the general form,

$$|\psi\rangle = (a(|\uparrow\uparrow\downarrow\rangle - |\downarrow\uparrow\uparrow\rangle) + b(|\uparrow\downarrow\uparrow\rangle - |\downarrow\uparrow\uparrow\rangle)) |uuu\rangle |C_A\rangle . \qquad (11.14)$$

However, this wave function cannot be made as totally antisymmetric for any a and b. This is the reason why spin-1/2 $|uuu\rangle$, $|ddd\rangle$, $|sss\rangle$ baryons do not exist.

11.4.2 Quark Structure of Proton

Proton is a spin-1/2 baryon made up of uud quarks. Since the spin-1/2 wave function cannot be symmetric nor antisymmetric, it may be expected from our theory that there is no spin-1/2 baryons. What was wrong with our theory?

Since the quark structure of the proton is a superposition of uud quarks, the quark structure of the proton is a superposition of $|uud\rangle$, $|udu\rangle$ and $|duu\rangle$. Since each quark has spin, actually the quark structure is a superposition of $|u(\Uparrow)u(\Uparrow)d(\Downarrow)\rangle$, $|u(\Uparrow)u(\Downarrow)d(\Uparrow)\rangle$, etc.

$$|\Psi\rangle = C_1 |u(\Uparrow)u(\Uparrow)d(\Downarrow)\rangle + C_2 |u(\Uparrow)u(\Downarrow)d(\Uparrow)\rangle + \cdots + C_9 |d(\Downarrow)u(\Uparrow)u(\Uparrow)\rangle .$$

$$(11.15)$$

It is possible to construct a transition equation with a 9×9 matrix and obtain the energy eigenstates as before. The wave function below is the totally antisymmetric mass eigenstate which corresponds to the wave function of the proton,

$$|p\rangle = \frac{1}{3\sqrt{2}} \begin{pmatrix} |udu\rangle \, (|\uparrow\uparrow\downarrow\rangle + |\downarrow\uparrow\uparrow\rangle - 2|\uparrow\downarrow\uparrow\rangle) \\ + |uud\rangle \, (|\uparrow\downarrow\uparrow\rangle + |\downarrow\uparrow\uparrow\rangle - 2|\uparrow\uparrow\downarrow\rangle) \\ + |duu\rangle \, (|\uparrow\downarrow\uparrow\rangle + |\uparrow\uparrow\downarrow\rangle - 2|\downarrow\uparrow\uparrow\rangle) \end{pmatrix} |C_A\rangle \, . \tag{11.16}$$

This wave function cannot be factorized into flavor part and spin part as

$$|p\rangle \neq (a_1 |uud\rangle + a_2 |udu\rangle + a_3 |duu\rangle)(b_1 |\uparrow\downarrow\uparrow\rangle + b_2 |\downarrow\uparrow\uparrow\rangle + b_3 |\uparrow\uparrow\downarrow\rangle) |C_A\rangle \, . \tag{11.17}$$

This means we cannot treat flavor and spin separately. We need to treat the flavor and spin as a whole. Mathematically speaking, we have to use $SU(6)$ group instead of $SU(3)_{\text{flavor}} \times SU(2)_{\text{spin}}$ to express the proton structure.

11.4.3 Λ, Σ^0 and Σ^{0*}

Both Λ and Σ^0 baryons consist of u, d, s quarks and have spin-1/2. But the masses are different; $m_\Lambda = 1116$ MeV, $m_{\Sigma^0} = 1193$ MeV. What is the structural difference between Λ and Σ^0? In the $|uds\rangle$ baryons, transition between $|su\rangle \leftrightarrow |us\rangle$ takes place by virtual K^+ exchange as shown in Fig. 11.6a. Since the mass of the K^+ is much heavier than π^+ mass, the amplitude of $|su\rangle \leftrightarrow |us\rangle$ transition is much smaller than that of $|du\rangle \leftrightarrow |ud\rangle$ transition shown in Fig. 11.6b. Therefore, we ignore the transition of s-quark. There is also the self-transition as shown in Fig. 11.7. By writing the wave function of uds baryon as

$$|\psi_{uds}\rangle = C_{ud} |ud\rangle |s\rangle + C_{du} |du\rangle |s\rangle \, . \tag{11.18}$$

We obtain the transition equation,

$$i\frac{d}{dt} \begin{pmatrix} C_{ud} \\ C_{du} \end{pmatrix} = \frac{g_\pi^2}{2} \begin{pmatrix} -1 & 2 \\ 2 & -1 \end{pmatrix} \begin{pmatrix} C_{ud} \\ C_{du} \end{pmatrix} \, . \tag{11.19}$$

The mass eigenstates are

$$\begin{cases} |\psi_+\rangle = \frac{1}{\sqrt{2}}(|ud\rangle |s\rangle + |du\rangle |s\rangle)e^{-i(M_0 + g_\pi^2/2)t} \\ |\psi_-\rangle = \frac{1}{\sqrt{2}}(|ud\rangle |s\rangle - |du\rangle |s\rangle)e^{-i(M_0 - 3g_\pi^2/2)t} \, , \end{cases} \tag{11.20}$$

where M_0 is the original $|uds\rangle$ mass ($\sim m_s + m_d + m_u$).

There are also $\Sigma^+(suu)$ and $\Sigma^-(sdd)$ baryons. The pion exchange self-transitions are taking place in them as shown in Fig. 11.8. The mass eigenstates are

$$\left|\Sigma^+\right\rangle = |uu\rangle |s\rangle \, e^{-i(M_0 + g_\pi^2/2)t}, \quad \left|\Sigma^-\right\rangle = |dd\rangle |s\rangle \, e^{-i(M_0 + g_\pi^2/2)t} \, . \tag{11.21}$$

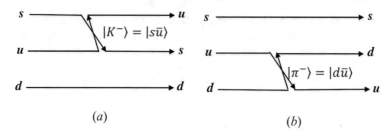

Fig. 11.6 a $|su\rangle \leftrightarrow |us\rangle$ transition. **b** $|du\rangle \leftrightarrow |ud\rangle$ transition

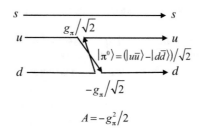

Fig. 11.7 $|sud\rangle \leftrightarrow |sud\rangle$ self-transition

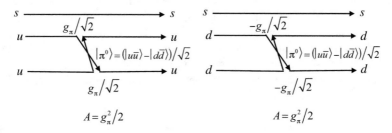

Fig. 11.8 $|suu\rangle \leftrightarrow |suu\rangle$, $|sdd\rangle \leftrightarrow |sdd\rangle$ self-transition

Therefore, the mass of ψ_+ and $\psi_{\Sigma^\pm}$ are the same and we can regard ψ_+ as ψ_{Σ^0}. ψ_- is considered to be another baryon with lighter mass called Λ baryon,

$$
\begin{cases}
|\Lambda\rangle = \frac{1}{\sqrt{2}}(|ud\rangle \, |s\rangle - |du\rangle \, |s\rangle)e^{-i(M_0 - 3g_\pi^2/2)t} \\
|\Sigma^0\rangle = \frac{1}{\sqrt{2}}(|ud\rangle \, |s\rangle + |du\rangle \, |s\rangle)e^{-i(M_0 + g_\pi^2/2)t} .
\end{cases}
\tag{11.22}
$$

From the wave functions (11.21) and (11.22), their masses are

$$
M_{\Sigma^0} = M_{\Sigma^+} = M_{\Sigma^-} = M_0 + g_\pi^2/2, \quad M_\Lambda = M_0 - 3g_\pi^2/2 .
\tag{11.23}
$$

The strength of the transition amplitude can be calculated from the mass difference of Σ^0 and Λ,

$$
g_\pi^2 = \frac{M_{\Sigma^0} - M_\Lambda}{2} \sim 40 \text{ MeV} .
\tag{11.24}
$$

We notice there is a similarity to the energy spread of the two spin system caused by the magnetic moment—magnetic moment (MM-MM) interactions.[2] There is a correspondence between quark flavors and spin directions as

$$|uu\rangle |s\rangle \to |\Uparrow\Uparrow\rangle, \quad |dd\rangle |s\rangle \to |\Downarrow\Downarrow\rangle,$$

$$\frac{(|ud\rangle + |du\rangle)) |s\rangle}{\sqrt{2}} \to \frac{(|\Uparrow\Downarrow\rangle + |\Downarrow\Uparrow\rangle))}{\sqrt{2}}, \quad \frac{(|ud\rangle - |du\rangle)) |s\rangle}{\sqrt{2}} \to \frac{(|\Uparrow\Downarrow\rangle - |\Downarrow\Uparrow\rangle))}{\sqrt{2}}.$$

$$(11.25)$$

We call $|u\rangle$ as Isospin-up state and $|d\rangle$ as Isospin-down state. Since the flavor structure of Σ^0 corresponds to the spin $|S = 1, S_z = 0\rangle$ state, we call Σ^0 as $I_z = 0$ component of the Isospin-1 state of $|uds\rangle$ baryon and we call Λ as Isospin-0 state of the $|uds\rangle$ system.

If we include the effect of $(|su\rangle \leftrightarrow |us\rangle)$ and $(|sd\rangle \leftrightarrow |ds\rangle)$ transitions and take into account the spin and color wave functions, the totally antisymmetric wave function of Λ baryon is

$$|\Lambda\rangle = \frac{1}{2\sqrt{3}} \begin{pmatrix} (|\Uparrow\Downarrow\Uparrow\rangle - |\Downarrow\Uparrow\Uparrow\rangle))(|uds\rangle - |dus\rangle)) \\ + (|\Uparrow\Uparrow\Downarrow\rangle - |\Downarrow\Uparrow\Uparrow\rangle))(|usd\rangle - |dsu\rangle)) \\ + (|\Uparrow\Uparrow\Downarrow\rangle - |\Uparrow\Downarrow\Uparrow\rangle))(|sud\rangle - |sdu\rangle)) \end{pmatrix} |C_A\rangle. \quad (11.26)$$

And that of Σ^0 baryon is

$$|\Sigma^0\rangle = \frac{1}{6} \begin{pmatrix} (2|\Uparrow\Uparrow\Downarrow\rangle - |\Uparrow\Downarrow\Uparrow\rangle - |\Downarrow\Uparrow\Uparrow\rangle))(|uds\rangle + |dus\rangle)) \\ + (2|\Downarrow\Uparrow\Uparrow\rangle - |\Uparrow\Downarrow\Uparrow\rangle - |\Uparrow\Uparrow\Downarrow\rangle))(|usd\rangle + |dsu\rangle)) \\ + (2|\Uparrow\Downarrow\Uparrow\rangle - |\Downarrow\Uparrow\Uparrow\rangle - |\Uparrow\Uparrow\Downarrow\rangle))(|sud\rangle + |sdu\rangle)) \end{pmatrix} |C_A\rangle. \quad (11.27)$$

On the other hand, the $|uds\rangle$ wave function of spin-3/2 and isospin-1, called Σ^{0*} baryon is

$$\left|\Sigma^{0*}\right\rangle = \frac{1}{\sqrt{6}} ((|uds\rangle + |dus\rangle) + (|usd\rangle + |dsu\rangle) + (|sud\rangle + |sdu\rangle)) |\Uparrow\Uparrow\Uparrow\rangle |C_A\rangle.$$

$$(11.28)$$

However, it is impossible to form totally antisymmetric $|uds\rangle$ wave function with spin-3/2 and isospin-0. This is the reason there is no Λ^* baryon in Table 11.1.

11.5 Isospin

The very basic process of the strong interactions in hadrons is the quark transitions emitting or absorbing the pions as shown in Fig. 11.9. If we write the quark state as

$$\left|\psi_q\right\rangle = c_u |u\rangle + c_d |d\rangle, \quad (11.29)$$

---

[2] See Chap. 3.

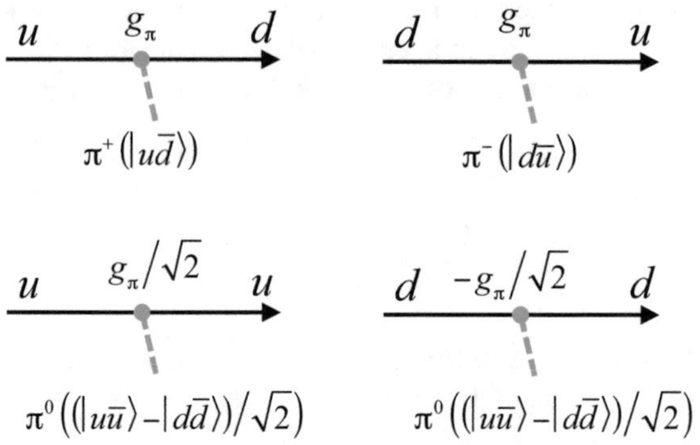

Fig. 11.9 Basic transitions
of u, d quarks

the transition equation corresponding to Fig. 11.9 is,

$$i\frac{d}{dt}\begin{pmatrix} c_u \\ c_d \end{pmatrix} = g_\pi \begin{pmatrix} 1/\sqrt{2} & 1 \\ 1 & -1/\sqrt{2} \end{pmatrix}\begin{pmatrix} c_u \\ c_d \end{pmatrix} \tag{11.30}$$

This form is exactly the same as the Pauli equation for the interactions between the
spin magnetic moment $\vec{\mu} = \mu\vec{\sigma}$ and the magnetic field $\vec{B}$ with $\mu\vec{B} = g_\pi(1, 0, 1/\sqrt{2})$.
This means that there is always correspondence between the strong interactions of
u, d-quarks and electromagnetic interactions of $|\Uparrow\rangle$ and $|\Downarrow\rangle$ spins,

$$|u\rangle \rightarrow |\Uparrow\rangle , \quad |d\rangle \rightarrow |\Downarrow\rangle . \tag{11.31}$$

That is the reason that we can understand the structure of the Σ and Λ baryon by
the concept of isospin.

As for the antiquarks, the general wave function can be written as

$$|\psi_{\bar{q}}\rangle = c_{\bar{u}}|\bar{u}\rangle + c_{\bar{d}}|\bar{d}\rangle . \tag{11.32}$$

The sign of the coupling constant for the strong interaction is negative of that of
quarks. Therefore, the transition equation for anti-quarks is

$$i\frac{d}{dt}\begin{pmatrix} c_{\bar{u}} \\ c_{\bar{d}} \end{pmatrix} = -g_\pi \begin{pmatrix} 1/\sqrt{2} & 1 \\ 1 & -1/\sqrt{2} \end{pmatrix}\begin{pmatrix} c_{\bar{u}} \\ c_{\bar{d}} \end{pmatrix} . \tag{11.33}$$

These equations can be rearranged as follows,

$$i\frac{d}{dt}\begin{pmatrix} -c_{\bar{d}} \\ c_{\bar{u}} \end{pmatrix} = g_\pi \begin{pmatrix} 1/\sqrt{2} & 1 \\ 1 & -1/\sqrt{2} \end{pmatrix}\begin{pmatrix} -c_{\bar{d}} \\ c_{\bar{u}} \end{pmatrix} . \tag{11.34}$$

If we compare this relation with Eq. (11.30), we notice that there is following correspondence,

$$|\Uparrow\rangle \rightarrow |u\rangle \rightarrow -|\bar{d}\rangle, \quad |\Downarrow\rangle \rightarrow |d\rangle \rightarrow |\bar{u}\rangle . \tag{11.35}$$

Therefore, $(-|\bar{d}\rangle)$ is the isospin-up state and $|\bar{u}\rangle$ is the isospin-down state. An effect of the minus sign of $|d\rangle$ has already been appeared in the quark structure of the π^0 system in Chap. 10 without the isospin concept. The isospin of π^0 is ($I = 1$, $I_z = 0$). Therefore, the quark isospin structure of π^0 is

$$|\pi^0\rangle = \frac{1}{\sqrt{2}}(|\Uparrow\Downarrow\rangle_I + |\Downarrow\Uparrow\rangle_I) = \frac{1}{\sqrt{2}}(|u\bar{u}\rangle + |d(-\bar{d})\rangle) = \frac{1}{\sqrt{2}}(|u\bar{u}\rangle - |d\bar{d}\rangle).$$

$$(11.36)$$

The minus sign of the superposition in the π^0 wave function comes from the minus sign of $(|-\bar{d}\rangle)$. Strictly speaking, the π^+ wave function has to be written as

$$|\pi^+\rangle = |\Uparrow\Uparrow\rangle_I = |u(-\bar{d})\rangle = -|u\bar{d}\rangle. \qquad (11.37)$$

However, in most cases the overall minus sign is omitted because it does not cause any physical effects.

Part V
Unknown Origin

Neutrino Oscillation: Relativistic Oscillation of Three-Flavor System

12

12.1 Introduction

Neutrino is a neutral fermion. In the standard model of the elementary particles, there are three kinds of neutrinos,

$$\nu_e, \quad \nu_\mu, \quad \nu_\tau . \tag{12.1}$$

The identity of these neutrinos is called "flavor". The neutrinos are assumed to be massless in the standard model. However, since neutrino oscillation was discovered in 1998, neutrinos are known to have small but finite masses because oscillations do not happen if they are massless. Therefore, we know now that the standard model needs to be expanded to include the neutrino masses.

The neutrino oscillation is a good example to learn about relativistic oscillation and mixing of three-flavor system. In this chapter, we start with two-flavor neutrino oscillation and proceed to the three-flavor oscillation which includes a much wider possibility than two flavor oscillations.

12.2 Two-Flavor Oscillation

Since neutrino mass is extremely small, the neutrinos are traveling with the relativistic speed in our experimental condition. We start the discussion with neutrino oscillation between ν_e and ν_μ at rest and then we apply Lorentz boost to the system to obtain the oscillation formula of the relativistic neutrinos.

© Springer Nature Switzerland AG 2021
F. Suekane, *Quantum Oscillations*, Lecture Notes in Physics 985,
https://doi.org/10.1007/978-3-030-70527-5_12

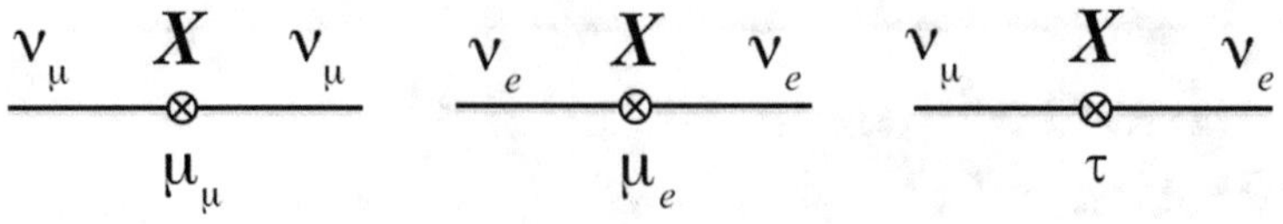

$$\nu_\mu \quad X \quad \nu_\mu \qquad \nu_e \quad X \quad \nu_e \qquad \nu_\mu \quad X \quad \nu_e$$
$$\mu_\mu \qquad\qquad \mu_e \qquad\qquad \tau$$

Fig. 12.1 Transitions between ν_e and ν_μ. We do not know what (X) causes these transitions

12.2.1 Neutrino Transition Amplitudes

We know something (X) gives (ν_e, ν_μ) the original masses (μ_e, μ_μ) and transforms the flavor between ν_e to ν_μ with transition amplitude τ as shown in Fig. 12.1. μ_e and μ_μ are real numbers and τ is generally a complex number. However, the imaginary phase does not physically appear in the two-flavor neutrino oscillation formula and we regard τ also as a real number to simplify the formulas. We do not know what X is yet. An important purpose of neutrino oscillation study is to investigate what X is based on the transition amplitudes measured by experiments.

Since there is cross-transition amplitude, the general wave function of the two neutrino system is a superposition of $|\nu_e\rangle$ and $|\nu_\mu\rangle$,

$$\psi_\nu[t] = C_\mu[t]\,|\nu_\mu\rangle + C_e[t]\,|\nu_e\rangle . \tag{12.2}$$

The transition equation of the neutrinos is

$$i\frac{d}{dt}\begin{pmatrix} C_\mu \\ C_e \end{pmatrix} = \begin{pmatrix} \mu_\mu & \tau \\ \tau & \mu_e \end{pmatrix}\begin{pmatrix} C_\mu \\ C_e \end{pmatrix} . \tag{12.3}$$

From the equation, the mass eigenstates are, from Eq. (B.5) as [1]

$$\begin{cases} |\psi_+[t]\rangle = (\cos\theta\,|\nu_\mu\rangle + \sin\theta\,|\nu_e\rangle)e^{-im_+t} \equiv |\nu_+\rangle\,e^{-im_+t} \\ |\psi_-[t]\rangle = (-\sin\theta\,|\nu_\mu\rangle + \cos\theta\,|\nu_e\rangle)e^{-im_-t} \equiv |\nu_-\rangle\,e^{-im_-t} . \end{cases} \tag{12.4}$$

The mixing angle is defined by the mixing triangle shown in Fig. 12.2 as

$$\tan 2\theta = \frac{2\tau}{\mu_\mu - \mu_e} = \frac{\tau}{\mu_-} . \tag{12.5}$$

And the masses are

$$m_\pm = \mu_+ \pm \omega , \tag{12.6}$$

where $\mu_\pm = \frac{\mu_\mu \pm \mu_e}{2}$ and $\omega = \sqrt{\mu_-^2 + \tau^2}$. Since $\omega \geq 0$, the masses in this expression have the relation $m_+ \geq m_-$. The relation between the basis states of mass eigenstates

---

[1] The arbitrary phases $\delta_\pm$ in Eq. (B.5) do not generate physical effect and $\delta_\pm = 0$ are assumed.

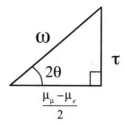

Fig. 12.2 Two-flavor neutrino mixing triangle. $\tan 2\theta = \frac{2\tau}{\mu_\mu - \mu_e}$, $\omega = \sqrt{\left(\frac{\mu_\mu - \mu_e}{2}\right)^2 + \tau^2}$

$|\nu_\pm\rangle$ and flavor eigenstates ($|\nu_e\rangle$, $|\nu_\mu\rangle$) is

$$\begin{pmatrix} |\nu_-\rangle \\ |\nu_+\rangle \end{pmatrix} = \begin{pmatrix} \cos\theta & -\sin\theta \\ \sin\theta & \cos\theta \end{pmatrix} \begin{pmatrix} |\nu_e\rangle \\ |\nu_\mu\rangle \end{pmatrix} . \tag{12.7}$$

12.2.2 Oscillation

The general wave function of the two-flavor neutrino system can be written using the mass eigenstate wave functions as

$$|\psi_\nu[t]\rangle = C_- |\nu_-\rangle e^{-im_- t} + C_+ |\nu_+\rangle e^{-im_+ t} . \tag{12.8}$$

If the state is pure $|\nu_\mu\rangle$ at $t = 0$, there is a relation,

$$\begin{aligned} |\psi_\mu[0]\rangle = |\nu_\mu\rangle &= C_- |\nu_-\rangle + C_+ |\nu_+\rangle \\ &= (C_- \cos\theta + C_+ \sin\theta) |\nu_e\rangle + (C_+ \cos\theta - C_- \sin\theta) |\nu_\mu\rangle . \end{aligned} \tag{12.9}$$

From the relation, $C_\pm$ are determined as

$$C_- = -\sin\theta, \quad C_+ = \cos\theta . \tag{12.10}$$

Therefore, the wave function at arbitrary time t is

$$|\psi_\nu[t]\rangle = \sin\theta \cos\theta (e^{-im_+ t} - e^{-im_- t}) |\nu_e\rangle + (\cos^2\theta e^{-im_+ t} + \sin^2\theta e^{-im_- t}) |\nu_\mu\rangle . \tag{12.11}$$

The probability that the system is in $|\nu_e\rangle$ state at time t is

$$P_{\nu_\mu \to \nu_e}[t] = |\langle \nu_e | \psi_\nu[t]\rangle|^2 = \sin^2 2\theta \sin^2\left[\frac{m_+ - m_-}{2} t\right] = \sin^2 2\theta \sin^2 \omega t . \tag{12.12}$$

The oscillation probability (12.12) can also be derived intuitively as follows. There are two processes for changing flavors from ν_μ to ν_e as shown in Fig. 12.3. One is that ν_μ propagate in time as ν_+ to become ν_e and another one is that ν_μ propagate

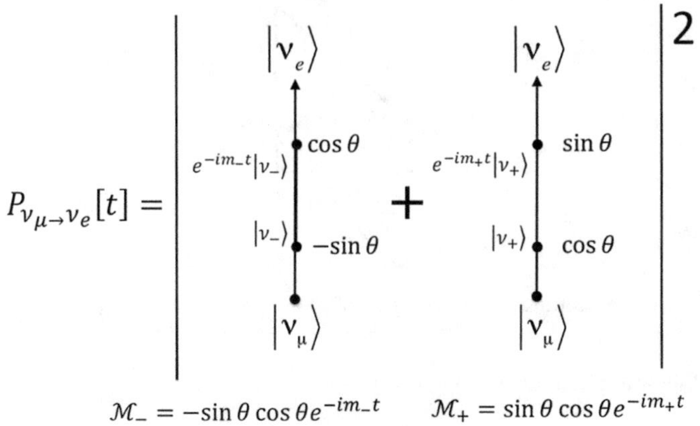

$$P_{\nu_\mu \to \nu_e}[t] = \left| \begin{array}{c} |\nu_e\rangle \\ e^{-im_-t}|\nu_-\rangle \uparrow \cos\theta \\ |\nu_-\rangle \downarrow -\sin\theta \\ |\nu_\mu\rangle \end{array} + \begin{array}{c} |\nu_e\rangle \\ e^{-im_+t}|\nu_+\rangle \uparrow \sin\theta \\ |\nu_+\rangle \downarrow \cos\theta \\ |\nu_\mu\rangle \end{array} \right|^2$$

$$\mathcal{M}_- = -\sin\theta\cos\theta e^{-im_-t} \qquad \mathcal{M}_+ = \sin\theta\cos\theta e^{-im_+t}$$

Fig. 12.3 The probability to start as ν_μ and change to ν_e at time t is calculated by taking the absolute square of the amplitudes of possible diagrams. In this case there are two diagrams that ν_μ changes to ν_e through ν_- or ν_+. The mass eigenstate ν_i acquires a phase e^{-im_it} after time t. The amplitude is the multiplication of this phase factor and ν_i component of ν_μ and ν_e component of ν_i

in time as ν_- to become ν_e. Since which process the neutrino undergoes can not be identifiable, quantum mechanical principle tells that the probability is the absolute square of the sum of the amplitudes of the two processes. The amplitude $\mathcal{M}_-$ is the product of $-\sin\theta$ (the amplitude of ν_- in ν_μ, see Eq. (12.7)) and e^{-im_-t} (phase acquired after ν_- propagate time t) and $\cos\theta$ (ν_e component in ν_-), that is

$$\mathcal{M}_- = -\cos\theta\sin\theta e^{-im_-t} . \tag{12.13}$$

Similarly,

$$\mathcal{M}_+ = \cos\theta\sin\theta e^{-im_+t} . \tag{12.14}$$

Therefore, the transition probability is

$$P_{\nu_\mu \to \nu_e} = |\mathcal{M}_- + \mathcal{M}_+|^2 = \left| -\sin\theta\cos\theta e^{-im_-t} + \sin\theta\cos\theta e^{-im_+t} \right|^2$$

$$= \sin^2 2\theta \sin^2\left[\frac{m_+ - m_-}{2}t\right] = \sin^2 2\theta \sin^2 \omega t , \tag{12.15}$$

which agrees with Eq. (12.12). This way of derivation of the probability is simple and intuitive and we use this method to derive the three-flavor oscillation in later sections.

12.2.3 Relativistic Oscillation Probability

Since the neutrino masses are very small, neutrinos are moving with relativistic speed in actual experimental conditions. To derive the oscillation formula of relativistic neutrinos, we start with a simple example. The transition equation of a particle with mass m at rest is

$$\frac{d}{dt}\psi[t] = -im\psi[t] \tag{12.16}$$

and its solution is

$$\psi[t] = e^{-imt} . \tag{12.17}$$

If we see the oscillation of neutrinos at rest from the frame (laboratory frame) which is moving with velocity $-v$ with respect to the direction of z axis, the relation between the space time coordinates of the laboratory frame; (t', z') and the neutrino frame (t, z) is expressed by the Lorentz transformation,

$$\begin{pmatrix} t' \\ z' \end{pmatrix} = \gamma \begin{pmatrix} 1 & \beta \\ \beta & 1 \end{pmatrix} \begin{pmatrix} t \\ z \end{pmatrix} : \quad \beta = \frac{v}{c}, \quad \gamma = \frac{1}{\sqrt{1-\beta^2}} , \tag{12.18}$$

where γ is the Lorentz factor. The space time coordinates of the neutrino in the neutrino frame are $(t, z) = (t, 0)$ and in the lab. frame are

$$(t', z') = (\gamma t, \gamma \beta t) = (t', \beta t') . \tag{12.19}$$

This means the neutrino is moving with velocity β as expected and the time on the neutrino frame is dilated by a factor γ in the lab. frame,

$$\psi[t] = e^{-imt} \xrightarrow{\text{L.T.}} \psi'[t', z'] = e^{-im\gamma(t'-\beta z')} = e^{-i(Et'-pz')} = e^{-ipx'} . \tag{12.20}$$

We call γm as energy E and $\gamma m \beta$ as momentum p.

The oscillation probability is calculated from the Fig. 12.4, as

$$P_{\nu_\mu \to \nu_e} = \sin^2\theta \cos^2\theta \left| e^{-ip_-x'} - e^{-ip_+x'} \right|^2 = \sin^2 2\theta \sin^2 \left[\frac{(p_+ - p_-)x'}{2} \right] . \tag{12.21}$$

The mixing angle is a function of the ratio of transition amplitudes and dimensionless and does not depend on the frame as indicated by Fig. 12.5.

In actual experiments, neutrinos are generated in pion or muon or beta decays,

$$\mathcal{X} \to \mathcal{Y} + \nu , \tag{12.22}$$

where $\mathcal{X}$ is the parent particle and $\mathcal{Y}$ can be single or multi particle system. In the $\mathcal{X}$ rest frame, the energy $(E_\pm)$ and momentum $(p_\pm)$ of a neutrino with mass $m_\pm$ are

$$E_\pm \sim E_0 \left(1 - \frac{m_\pm^2}{2M_\mathcal{X} E_0} \right), \quad p_\pm \sim E_0 \left(1 - \frac{m_\pm^2}{2E_0^2} \left(1 + \frac{E_0}{M_\mathcal{X}} \right) \right), \tag{12.23}$$

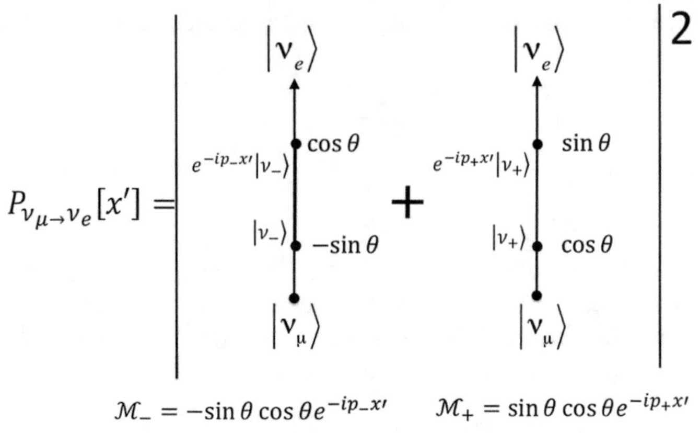

$$P_{\nu_\mu \to \nu_e}[x'] = \left| \begin{array}{cc} \begin{array}{c} |v_e\rangle \\ \\ e^{-ip_- x'}|v_-\rangle \uparrow \cos\theta \\ \\ |v_-\rangle \downarrow -\sin\theta \\ \\ |v_\mu\rangle \end{array} & + & \begin{array}{c} |v_e\rangle \\ \\ e^{-ip_+ x'}|v_+\rangle \uparrow \sin\theta \\ \\ |v_+\rangle \downarrow \cos\theta \\ \\ |v_\mu\rangle \end{array} \end{array} \right|^2$$

$$\mathcal{M}_- = -\sin\theta\cos\theta e^{-ip_- x'} \qquad \mathcal{M}_+ = \sin\theta\cos\theta e^{-ip_+ x'}$$

Fig. 12.4 The probability to start as ν_μ and change to ν_e at space time x' is calculated by taking absolute square of the amplitudes of possible diagrams as in Fig. 12.3. In this case neutrino is moving with a relativistic speed. The mixing angle is a function of the ratio of transition amplitudes and dimensionless and does not depend on the frame (see Fig. 12.5)

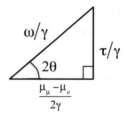

Fig. 12.5 Mixing triangle for relativistic neutrinos

where E_0 is the neutrino energy in case the neutrino mass is 0,

$$E_0 = \frac{M_\mathcal{X}^2 - M_\mathcal{Y}^2}{2M_\mathcal{X}}. \tag{12.24}$$

Usually, we measure neutrino oscillations as a function of the position z' rather than t'. Therefore, we express the oscillation phase in Eq. (12.21) as a function of z', as

$$\Phi = \frac{1}{2}((E_+ - E_-)t' - (p_+ - p_-)z') \sim \frac{m_+^2 - m_-^2}{4E_0}z', \tag{12.25}$$

where $z' = \beta t' \sim t'$ is used. Finally the oscillation probability with relativistic neutrino is

$$P_{\nu_\mu \to \nu_e} = \sin^2 2\theta \sin^2\left[\frac{\Delta m^2}{4E}z'\right]; \quad \Delta m^2 = m_+^2 - m_-^2. \tag{12.26}$$

This formula is often used in studies of neutrino oscillation. Note that this formula does not depend on the $\mathcal{Y}$ system. It means we do not need care about how neutrino is generated.

12.2.4 Another Way to Derive Relativistic Neutrino Oscillation

Since the neutrino position z' at time t' is $z' = \beta t'$, the wave function on the neutrino is

$$\psi'[t', z' = \beta t'] = e^{-i(m/\gamma)t'} . \tag{12.27}$$

Therefore, the differential equation for the time of the moving particles is expressed as

$$\frac{d}{dt'}\psi'[t'] = -i\frac{m}{\gamma}\psi'[t'] . \tag{12.28}$$

This means that the relativistic wave function of the particle can be obtained by replacing $m \to m' = m/\gamma$. Similarly, the relativistic two-flavor transition equation can be obtained by dividing all the transition amplitudes in Eq. (12.3) by γ as

$$i\frac{d}{dt'}\begin{pmatrix} C'_e \\ C'_\mu \end{pmatrix} = \frac{1}{\gamma}\begin{pmatrix} \mu_e & \tau \\ \tau & \mu_\mu \end{pmatrix}\begin{pmatrix} C'_e \\ C'_\mu \end{pmatrix} . \tag{12.29}$$

The mixing triangle of neutrino at rest (Fig. 12.2) is changed to Fig. 12.5 accordingly. Since the change is only the scale of the triangle, the mixing angle θ does not change. The oscillation probability for a relativistic neutrino becomes from Eq. (12.12) to

$$P'_{\nu_\mu \to \nu_e}[t'] = \sin^2 2\theta \sin^2 \frac{\omega}{\gamma}t' . \tag{12.30}$$

In the lab. frame, the energy of the mass eigenstate neutrinos is

$$E_\pm = \gamma m_\pm . \tag{12.31}$$

We use γ as the ratio of the average energy and average mass,

$$\gamma = \frac{\overline{E}}{\overline{m}} = \frac{2\overline{E}}{m_+ + m_-} . \tag{12.32}$$

Then the oscillation probability is

$$P'_{\nu_\mu \to \nu_e}[z'] = \sin^2 2\theta \sin^2 \frac{m_+^2 - m_-^2}{4\overline{E}}z' . \tag{12.33}$$

12.2.5 A Relation Between Transition Amplitudes and Neutrino Flavor Mass

In two-flavor neutrino case, there are three neutrino transition amplitudes (μ_e, μ_μ, τ) while only two parameters (θ, Δm^2) can be measured in neutrino oscillation experiments. Therefore, additional information is necessary to determine the neutrino transition amplitudes other than the neutrino oscillation experiments. Direct neutrino mass measurements can provide such missing information.

In direct neutrino mass measurements, the mass of the flavor eigenstate neutrino, such as ν_e, produced by the weak interactions is measured. However, the flavor eigenstate does not have fixed mass. If we measure the ν_e mass; m_{ν_e}, we will observe m_- with probability $\cos^2 \theta$ and m_+ with probability $\sin^2 \theta$. If the mass resolution of the experiment is not good enough to separate m_- and m_+, the mass measured in the experiments is the weighted average of m_- and m_+;

$$m_{\nu_e} = \cos^2 \theta m_- + \sin^2 \theta m_+ \ . \tag{12.34}$$

From the relations between the transition amplitudes and oscillation parameters, Eqs. (12.5) and (12.6), it can be shown that

$$m_{\nu_e} = \mu_+ - w \cos 2\theta = \mu_e \ . \tag{12.35}$$

This relation indicates that the effective mass of flavor eigenstate directly corresponds to the self-transition amplitude.[2] By combining ν_e mass, oscillation parameters; θ and Δm^2, it is possible to determine the transition amplitudes as follows:

$$\mu_e = m_{\nu_e}, \quad \mu_\mu = \sqrt{m_{\nu_e}^2 + \Delta m^2 \cos 2\theta},$$

$$\tau = \frac{1}{2} \left(\sqrt{m_{\nu_e}^2 + \Delta m^2 \cos 2\theta} - m_{\nu_e} \right) \tan 2\theta \ . \tag{12.36}$$

Note that in our definition, $\Delta m^2 \geq 0$ and the neutrino oscillation experiment measures $\sin^2 2\theta$ and the sign of $\cos 2\theta$ cannot be measured.

12.3   Three-Flavor Neutrino Oscillation

In the standard model, there are three kinds of neutrinos, Eq. (12.1). By generalizing the two-flavor case Eq. (12.2), we can write the general neutrino state at rest as superposition of flavor eigenstates as

$$\psi_\nu[t] = C_e[t] |\nu_e\rangle + C_\mu[t] |\nu_\mu\rangle + C_\tau[t] |\nu_\tau\rangle \tag{12.37}$$

and the corresponding transition equation (12.3) can be expressed as the following form:

$$\frac{d}{dt}\psi_\nu[t] = -iT\psi_\nu[t] \ . \tag{12.38}$$

From the conservation of probability,

$$|\psi_\nu[t]|^2 = |\psi_\nu[t + \delta t]|^2 = \left| \psi_\nu[t] + \dot{\psi}_\nu[t]\delta t \right|^2 = |\psi_\nu - iT\psi_\nu|^2$$
$$= |\psi_\nu[t]|^2 + i\{\psi_\nu^\dagger(T - T^\dagger)\psi_\nu\}\delta t + O[(\delta t)^2] \ . \tag{12.39}$$

---

[2]In actual experiments, $m_{\nu_e}^2 = \cos^2 \theta m_-^2 + \sin^2 \theta m_+^2 = \frac{3\mu_e^2 + \mu_\mu^2}{4} + \tau^2$ is measured.

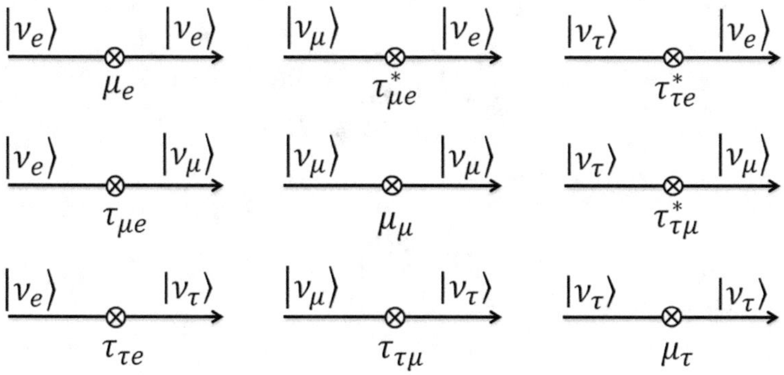

Fig. 12.6 Transitions among ν_e, ν_μ and ν_τ

Therefore, the transition matrix has to be Hermitian,

$$T^\dagger = T \ . \tag{12.40}$$

This means that the 9 transition amplitudes can be written as shown in Fig. 12.6. The transition equation for relativistic three-flavor neutrino can be written, from analogy of Eq. (12.29), as

$$\frac{d}{dt}\begin{pmatrix} C_e \\ C_\mu \\ C_\tau \end{pmatrix} = -\frac{i}{\gamma}\begin{pmatrix} \mu_e & \tau^*_{\mu e} & \tau^*_{\tau e} \\ \tau_{\mu e} & \mu_\mu & \tau^*_{\tau \mu} \\ \tau_{\tau e} & \tau_{\tau \mu} & \mu_\tau \end{pmatrix}\begin{pmatrix} C_e \\ C_\mu \\ C_\tau \end{pmatrix} \equiv -\frac{i}{\gamma}T\Psi \ . \tag{12.41}$$

By expressing the three mass eigenstates as

$$|\nu_i\rangle \, e^{-i(m_i/\gamma)t} \ : \quad \{i = 1 \sim 3\} \ , \tag{12.42}$$

the wave function (12.37) can also be expressed by the mass eigenstates,

$$\psi_\nu[x] = C_1[0]\,|\nu_1\rangle \, e^{-i(m_1/\gamma)t} + C_2[0]\,|\nu_2\rangle \, e^{-i(m_2/\gamma)t} + C_3[0]\,|\nu_3\rangle \, e^{-i(m_3/\gamma)t} \ . \tag{12.43}$$

We express the coefficients by matrices as follows:

$$\Psi = \begin{pmatrix} C_e[t] \\ C_\mu[t] \\ C_\tau[t] \end{pmatrix}, \quad \Phi = \begin{pmatrix} C_1[t] \\ C_2[t] \\ C_3[t] \end{pmatrix} \ . \tag{12.44}$$

The transition equation (12.41) and the transition equation of the mass eigenstate can be expressed as

$$\dot{\Psi} = -(i/\gamma)T\Psi \ , \quad \dot{\Phi} = -(i/\gamma)M\Phi \ , \tag{12.45}$$

where

$$M = \begin{pmatrix} m_1 & 0 & 0 \\ 0 & m_2 & 0 \\ 0 & 0 & m_3 \end{pmatrix} .$$

(12.46)

Now we relate Ψ and Φ by a matrix U,

$$\Phi = U\Psi .$$

(12.47)

Then the first equation in Eq. (12.45) becomes

$$\dot{\Phi} = -(i/\gamma)UTU^{-1}\Phi .$$

(12.48)

By comparing to the second equation in Eq. (12.45), if the matrix U satisfies

$$UTU^{-1} = M ,$$

(12.49)

Equation (12.47) is satisfied. The mass eigenstate is obtained by Eq. (12.47) and the masses are obtained by Eq. (12.49). Furthermore, if we take the complex conjugate of Eq. (12.49),

$$U^{-1^\dagger}TU^\dagger = M ,$$

(12.50)

where $T^\dagger = T$ and real diagonal M are are used. By comparing with Eq. (12.49), we can show that the matrix U is unitary ($U^\dagger = U^{-1}$). We write this U as U_ν to show this is the neutrino mixing matrix. U_ν is called Pontercorvo-Maki-Nakagawa-Sakata (PMNS) matrix. The relation between mass and flavor eigenstates is

$$\begin{pmatrix} |\nu_1\rangle \\ |\nu_2\rangle \\ |\nu_3\rangle \end{pmatrix} = \begin{pmatrix} U_{e1} & U_{\mu 1} & U_{\tau 1} \\ U_{e2} & U_{\mu 2} & U_{\tau 2} \\ U_{e3} & U_{\mu 3} & U_{\tau 3} \end{pmatrix} \begin{pmatrix} |\nu_e\rangle \\ |\nu_\mu\rangle \\ |\nu_\tau\rangle \end{pmatrix} = U_\nu^T \begin{pmatrix} |\nu_e\rangle \\ |\nu_\mu\rangle \\ |\nu_\tau\rangle \end{pmatrix} .$$

(12.51)

There are some differences from the two-flavor oscillation case. For three-flavor case, the masses, m_i and mixing matrix elements, $U_{\alpha i}$ consist of hundreds of terms from the combinations of the transition amplitudes, μ_α, $\tau_{\alpha\beta}$. However, some simple relations still hold. For example,

$$m_1 + m_2 + m_3 = \mu_e + \mu_\mu + \mu_\tau ,$$
$$m_{\nu_\alpha} = |U_{\alpha 1}|^2 m_1 + |U_{\alpha 2}|^2 m_2 + |U_{\alpha 3}|^2 m_3 = \mu_\alpha ,$$

(12.52)

where m_{ν_α} is the flavor mass.

Another difference is that an imaginary phase remains in the oscillation probability and it can generate CP violation effect.

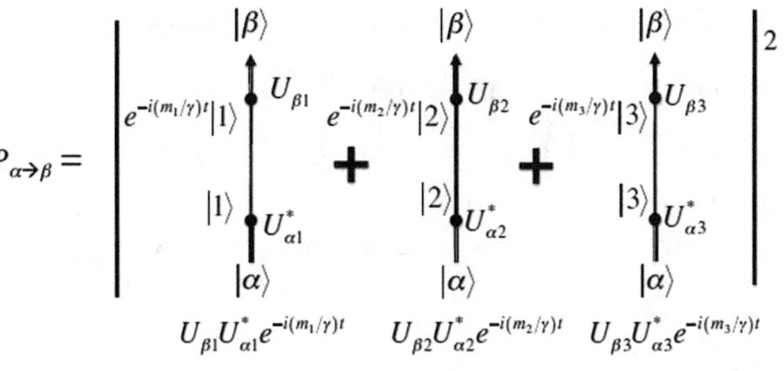

Fig. 12.7 Transition from ν_α to ν_β

The oscillation probabilities can be calculated using Fig. 12.7. The probability of the transition from ν_α to ν_β can be calculated as[3]

$$
P_{\alpha \to \beta} = \left| \sum_i U_{\beta i} U_{\alpha i}^* e^{-i(m_i/\gamma)t} \right|^2 = \cdots
$$
$$
= \delta_{\alpha\beta} - 4 \sum_{i>j} \sin^2 \Phi_{ij} \Re \left[\Lambda_{ij}^{\alpha\beta} \right] + 2 \sum_{i>j} \sin 2\Phi_{ij} \Im \left[\Lambda_{ij}^{\alpha\beta} \right] , \tag{12.53}
$$

where

$$
\Lambda_{ij}^{\alpha\beta} \equiv U_{\alpha i} U_{\beta i}^* U_{\alpha j}^* U_{\beta j}, \quad \Phi_{ij} = \frac{\Delta m_{ij}^2}{4E} L . \tag{12.54}
$$

Assuming the CPT invariance, we can relate the oscillation probability of antineutrinos to that of neutrinos,

$$
P_{\bar\beta \to \bar\alpha} = P_{\alpha \to \beta} = P_{\beta \to \alpha}^* = \delta_{\alpha\beta} - 4 \sum_{i>j} \sin^2 \Phi_{ij} \Re \left[\Lambda_{ij}^{\alpha\beta} \right] - 2 \sum_{i>j} \sin 2\Phi_{ij} \Im \left[\Lambda_{ij}^{\alpha\beta} \right] , \tag{12.55}
$$

where the first equation comes from the CPT invariance and the second equation comes from the relation $\Lambda_{kj}^{\beta\alpha} = \left(\Lambda_{kj}^{\alpha\beta} \right)^*$.

In general, a 3×3 unitary matrix can include three imaginary phases. However, two of them can be absorbed in the wave functions without affecting physics and one imaginary phase remain. There are three real and one imaginary free parameters in the neutrino mixing matrix and in order to interpret the various experimental results,

---

[3]A complete derivation can be found in Ref. [1].

it is convenient to parametrize the mixing matrix as shown below,

$$
\begin{aligned}
U_\nu &= \begin{pmatrix} 1 & 0 & 0 \\ 0 & c_{23} & s_{23} \\ 0 & -s_{23} & c_{23} \end{pmatrix} \begin{pmatrix} c_{13} & 0 & s_{13}e^{-i\delta} \\ 0 & 1 & 0 \\ -s_{13}e^{i\delta} & 0 & c_{13} \end{pmatrix} \begin{pmatrix} s_{12} & s_{12} & 0 \\ -s_{12} & c_{12} & 0 \\ 0 & 0 & 1 \end{pmatrix} \\
&= \begin{pmatrix} c_{12}c_{13} & s_{12}c_{13} & s_{13}e^{-i\delta} \\ -s_{12}c_{23} - c_{12}s_{23}s_{13}e^{i\delta} & c_{12}c_{23} - s_{12}s_{23}s_{13}e^{i\delta} & s_{23}c_{13} \\ s_{12}s_{23} - c_{12}c_{23}s_{13}e^{i\delta} & -s_{23}c_{12} - s_{12}c_{23}s_{13}e^{i\delta} & c_{23}c_{13} \end{pmatrix},
\end{aligned}
\tag{12.56}
$$

where $s_{ij} = \sin\theta_{ij}$, $c_{ij} = \cos\theta_{ij}$.

12.4   Measurements of Oscillation Parameters

There have been a number of neutrino oscillation experiments and all the mixing angles, θ_{12}, θ_{23}, θ_{13} and two mass square differences Δm_{12}^2, Δm_{13}^2 have been measured. Figure 12.8 summarizes the baseline and neutrino energy relation for such experiments.

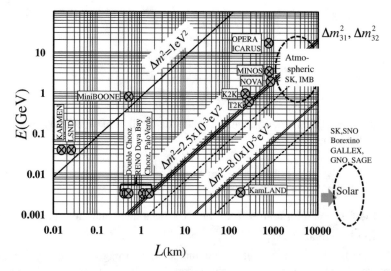

Fig. 12.8 Relation between energy (E) and baseline (L) of various neutrino oscillation experiments which observed positive results. The solid lines are the first oscillation maximum. The dashed lines are the second oscillation maximum. The oscillation on $\Delta m^2 = 1$ eV^2 is not established

12.4.1 θ_{23} and Δm^2_{23}

The probability of $\nu_\mu \to \nu_\mu$ disappearance at $E/L \sim \Delta m^2_{23}/2\pi$ is expressed as

$$P_{\nu_\mu \to \nu_\mu}[L] \sim 1 - \sin^2 2\theta_{23} \sin^2 \frac{\Delta m^2_{32}}{4E} L . \qquad (12.57)$$

Kamiokande, Super Kamiokande, T2K, MINOS, NOVA experiments measured this oscillation mode using atmospheric and accelerator produced neutrinos. They have measured

$$\sin^2 2\theta_{23} \sim 1, \quad \Delta m^2_{23} \sim 2.5 \times 10^{-3} \text{ eV}^2. \qquad (12.58)$$

This is almost a full mixing.

12.4.2 θ_{12} and Δm^2_{12}

θ_{12} and Δm^2_{12} were measured by ν_e and $\bar{\nu}_e$ disappearance at $E/L \sim \Delta m^2_{12}/2\pi$,

$$P_{\nu_e \to \nu_e} \sim 1 - \sin^2 2\theta_{12} \sin^2 \left[\frac{\Delta m^2_{12}}{4E} L \right] . \qquad (12.59)$$

Solar neutrino can satisfy this condition and Homestake, Super Kamiokande, SAGE, GALLEX, GNO, SNO, and Borexino measured this oscillation. In addition to the oscillation parameter measurement, $m_+ > m_-$ was also determined by using matter effect of the solar neutrino. KamLAND experiment measured reactor neutrino deficit at $L \sim 180$ km and measured clear oscillation pattern as shown in Fig. 12.9.

The solar neutrino and reactor neutrino experiments show

$$\sin^2 2\theta_{12} \sim 0.85, \quad \Delta m^2_{12} \sim 7 \times 10^{-5} \text{ eV}^2 . \qquad (12.60)$$

Fig. 12.9 Neutrino oscillation pattern measure by KamLAND reactor neutrino experiment. From [2]

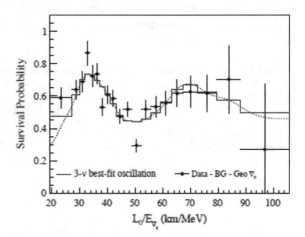

12.4.3 θ_{13}

The mixing angle θ_{13} can be measured by reactor neutrinos at a baseline $L \sim 1.3$ km,

$$P_{\nu_e \to \nu_e} \sim 1 - \sin^2 2\theta_{13} \sin^2 \left[\frac{\Delta m_{13}^2}{4E} L \right] . \tag{12.61}$$

Daya Bay, RENO and Double Chooz experiment have measured it to be

$$\sin^2 2\theta_{13} \sim 0.1 . \tag{12.62}$$

12.4.4 δ

The CP violation effect and the CP violation parameter δ can be measured by the asymmetry between $P_{\nu_\mu \to \nu_e} (\equiv P)$ and $P_{\bar{\nu}_\mu \to \bar{\nu}_e} (\equiv \overline{P})$ at $E/L \sim \Delta m_{13}^2 / 2\pi$,

$$A_{CP} = \frac{P - \overline{P}}{P + \overline{P}} \sim - \left| \frac{\Delta m_{12}^2}{\Delta m_{13}^2} \right| \frac{\pi \sin 2\theta_{12}}{\tan \theta_{23} \sin 2\theta_{13}} \sin \delta \sim -0.27 \sin \delta . \tag{12.63}$$

T2K and NOVA experiments measured this asymmetry and is showing an indication of

$$\sin \delta \sim -1 \tag{12.64}$$

using θ_{13} measured by the reactor neutrino experiments.

12.4.5 Summary of the Measurements

The oscillation parameter measured in each experiment is usually analyzed using the two-flavor oscillation formula. To determine the three-flavor oscillation parameters, a global analysis of all the data is necessary. Some groups perform such analysis and Table 12.1 shows such result.

Table 12.1 Summary of measurements. N.H. is the normal mass hierarchy ($m_3 > m_1$) case and I.H. is the inverted mass hierarchy ($m_3 < m_1$) case. Taken from Nufit5.0 [3]

	θ_{12}	θ_{23}	θ_{13}	δ_{CP}	Δm_{12}^2 [eV^2]	Δm_{31}^2 [eV^2]
N.H.	33.4°	49.2°	8.57°	195°	7.42×10^{-5}	2.51×10^{-3}
I.H.	33.6°	49.1°	8.55°	197°	7.42×10^{-5}	-2.52×10^{-3}

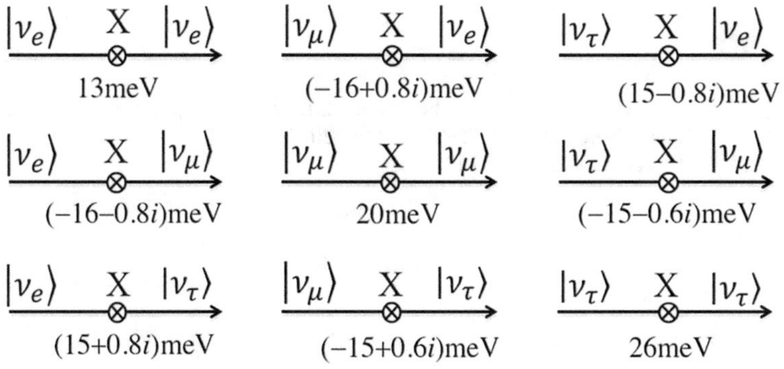

$$\underset{\underset{13\text{meV}}{\otimes}}{|\nu_e\rangle \;\; X \;\; |\nu_e\rangle} \qquad \underset{\underset{(-16+0.8i)\text{meV}}{\otimes}}{|\nu_\mu\rangle \;\; X \;\; |\nu_e\rangle} \qquad \underset{\underset{(15-0.8i)\text{meV}}{\otimes}}{|\nu_\tau\rangle \;\; X \;\; |\nu_e\rangle}$$

$$\underset{\underset{(-16-0.8i)\text{meV}}{\otimes}}{|\nu_e\rangle \;\; X \;\; |\nu_\mu\rangle} \qquad \underset{\underset{20\text{meV}}{\otimes}}{|\nu_\mu\rangle \;\; X \;\; |\nu_\mu\rangle} \qquad \underset{\underset{(-15-0.6i)\text{meV}}{\otimes}}{|\nu_\tau\rangle \;\; X \;\; |\nu_\mu\rangle}$$

$$\underset{\underset{(15+0.8i)\text{meV}}{\otimes}}{|\nu_e\rangle \;\; X \;\; |\nu_\tau\rangle} \qquad \underset{\underset{(-15+0.6i)\text{meV}}{\otimes}}{|\nu_\mu\rangle \;\; X \;\; |\nu_\tau\rangle} \qquad \underset{\underset{26\text{meV}}{\otimes}}{|\nu_\tau\rangle \;\; X \;\; |\nu_\tau\rangle}$$

Fig. 12.10 Transition amplitudes for the case $m_3 > m_2 > m_1 \sim 0$. "X" means the origin of the transitions is not yet understood

The mixing matrix depends on the yet undetermined mass hierarchy.[4] For N.H. case,

$$U_\nu[NH] = \begin{pmatrix} 0.83 & 0.54 & -0.14 + 0.04i \\ -0.27 + 0.02i & 0.61 + 0.02i & 0.75 \\ 0.50 + 0.02i & -0.58 + 0.01i & 0.65 \end{pmatrix}. \qquad (12.65)$$

The transition amplitude cannot be calculated because the data of the absolute neutrino mass has not been measured. However, if we assume $m_3 > m_2 > m_1 \sim 0$, The neutrino masses become

$$M = \begin{pmatrix} 0 & 0 & 0 \\ 0 & 8.6 & 0 \\ 0 & 0 & 51 \end{pmatrix} \text{meV} . \qquad (12.66)$$

and from the analogy of Eq. (8.65), the transition amplitudes are calculated as

$$T[m_3 > m_2 > m_1 \sim 0] \sim \begin{pmatrix} 13 & -16 + 0.8i & 15 - 0.8i \\ -16 - 0.8i & 20 & -15 - 0.6i \\ 15 + 0.8i & -15 + 0.6i & 26 \end{pmatrix} \text{meV} .$$

$$(12.67)$$

Figure 12.10 shows the transition amplitudes pictorially.

These transitions are very much smaller compared with the quark transition amplitudes Eq. (8.67) and it is regarded as unnatural to think that neutrino transition amplitudes are originated from the same Higgs mechanism.

---

[4]The relation of the size of m_1 and m_3. If $m_1 < m_3$, it is called normal mass hierarchy (N.H.) and if $m_1 > m_3$, it is called inverted mass hierarchy (I.H.).

References

1. Suekane, F.: Neutrino Oscillations. A Practical Guide to Basics and Applications. Springer (2015)
2. KamLAND collaboration, Phys. Rev. **D88**, 033001 (2013)
3. Esteban, I. et al.: arXiv:2007.14792v1 (2020)

Appendix A: Summary of Parameters and Formulas

A.1 Physics Parameters

$$c = 2.997\,924\,58 \times 10^8 \ (\text{m/s})$$
$$\hbar = 6.582\,119\,3 \times 10^{-22} \ (\text{MeV} \cdot \text{s})$$
$$\hbar c = 197.327 \ (\text{MeV} \cdot \text{fm})$$
$$\alpha = 1/137.0359990$$
$$G_F = 1.166\,378 \times 10^{-5} \ (1/\text{GeV}^2)(\text{How to memorize} : 1.08^2 = 1.1664)$$
$$\sin^2\theta_W^{\overline{MS}}(M_Z) = 0.23121$$
$$m_e = 0.510\,999 \ (\text{MeV/c}^2), \quad m_\mu = 105.653 \ (\text{MeV/c}^2), \tag{A.1}$$
$$m_\tau = 1776.86 \ (\text{MeV/c}^2)$$
$$m_p = 938.272 \ (\text{MeV/c}^2), \quad m_n = 939.565 \ (\text{MeV/c}^2)$$
$$m_Z = 91.188 \ (\text{GeV/c}^2), \quad m_W = 80.38 \ (\text{GeV/c}^2)$$
$$\mu_B = e\hbar/2m_e = 5.788\,381 \times 10^{-11} \ (\text{MeV/T}),$$
$$\mu_N = e\hbar/2m_p = 3.152\,451 \times 10^{-14} \ (\text{MeV/T})$$
$$a_B = 0.529\,117 \times 10^{-10} \ (\text{m})$$

A.2 Spinors

Two component spinor.

$$\chi_1 = \begin{pmatrix} 1 \\ 0 \end{pmatrix}, \quad \chi_2 = \begin{pmatrix} 0 \\ 1 \end{pmatrix} \tag{A.2}$$

© Springer Nature Switzerland AG 2021
F. Suekane, *Quantum Oscillations*, Lecture Notes in Physics 985,
https://doi.org/10.1007/978-3-030-70527-5_A

Spin wave function.

$$\hat{s}[\theta, \phi] = e^{i\delta} \begin{pmatrix} e^{-i(\phi/2)} \cos(\theta/2) \\ e^{i(\phi/2)} \sin(\theta/2) \end{pmatrix} \tag{A.3}$$

Spin direction vector.

$$\vec{e}_s = \{\hat{s}^\dagger[\theta, \phi]\boldsymbol{\sigma}\hat{s}[\theta, \phi]\} = \sin\theta \cos\phi \vec{e}_x + \sin\theta \sin\phi \vec{e}_y + \cos\theta \vec{e}_z \tag{A.4}$$

A.3   Pauli Matrix

Pauli matrices.

$$I = \begin{pmatrix} 1 & 0 \\ 0 & 1 \end{pmatrix}, \quad \sigma_x = \begin{pmatrix} 0 & 1 \\ 1 & 0 \end{pmatrix}, \quad \sigma_y = \begin{pmatrix} 0 & -i \\ i & 0 \end{pmatrix}, \quad \sigma_z = \begin{pmatrix} 1 & 0 \\ 0 & -1 \end{pmatrix} \tag{A.5}$$

Pauli matrix relations

$$\sigma_i \sigma_j = i\sigma_k; \quad (i, j, k) \text{ are cyclic.}$$
$$\sigma_i \sigma_j + \sigma_j \sigma_i = 2\delta_{ij} I \tag{A.6}$$

Pauli vector.

$$\vec{\sigma} = \begin{pmatrix} \vec{e}_z & \vec{e}_x - i\vec{e}_y \\ \vec{e}_x + i\vec{e}_y & -\vec{e}_z \end{pmatrix} = \begin{pmatrix} \vec{e}_z & \vec{e}_- \\ \vec{e}_+ & -\vec{e}_z \end{pmatrix} \tag{A.7}$$

Relation Pauli vector elements.

$$\vec{e}_\pm \cdot \vec{e}_\pm = 0, \quad \vec{e}_\pm \cdot \vec{e}_\mp = 2, \quad \vec{e}_\pm \cdot \vec{e}_z = 0. \tag{A.8}$$

Spin-magnetic field interaction.

$$\vec{B} \cdot \vec{\sigma} = \begin{pmatrix} B_z & B_x - iB_y \\ B_x + iB_y & -B_z \end{pmatrix} = \begin{pmatrix} B_z & B_- \\ B_+ & -B_z \end{pmatrix} \tag{A.9}$$

A.4   Dirac Matrix

Dirac matrices.

$$\gamma_0 = \begin{pmatrix} I & 0 \\ 0 & -I \end{pmatrix}, \quad \gamma_1 = \begin{pmatrix} 0 & \sigma_x \\ -\sigma_x & 0 \end{pmatrix}, \quad \gamma_2 = \begin{pmatrix} 0 & \sigma_y \\ -\sigma_y & 0 \end{pmatrix}, \quad \gamma_3 = \begin{pmatrix} 0 & \sigma_z \\ -\sigma_z & 0 \end{pmatrix} \tag{A.10}$$

Relations of Dirac matrices.

$$\gamma^\mu \gamma^\nu + \gamma^\nu \gamma^\mu = 2g^{\mu\nu}, \quad g^{\mu\nu} = \begin{pmatrix} 1 & 0 & 0 & 0 \\ 0 & -1 & 0 & 0 \\ 0 & 0 & -1 & 0 \\ 0 & 0 & 0 & -1 \end{pmatrix} \tag{A.11}$$

$$\gamma_5 = i\gamma_0\gamma_1\gamma_2\gamma_3 = \begin{pmatrix} 0 & I \\ I & 0 \end{pmatrix} \tag{A.12}$$

Chirality projection matrices.

$$\gamma_R = \frac{1+\gamma_5}{2} = \frac{1}{2}\begin{pmatrix} I & I \\ I & I \end{pmatrix}, \quad \gamma_L = \frac{1-\gamma_5}{2} = \frac{1}{2}\begin{pmatrix} I & -I \\ -I & I \end{pmatrix} \tag{A.13}$$

A.5   Dirac Equation

Dirac equation.

$$\left(i\gamma_\mu\partial^\mu - m\right)\psi[x] = 0 \tag{A.14}$$

General solution.

$$\psi_+[x] = \begin{pmatrix} u \\ (\vec{\eta}\cdot\vec{\sigma})u \end{pmatrix} e^{-i(Et-\vec{p}\vec{x})}, \quad \psi_-[x] = \begin{pmatrix} (\vec{\eta}\cdot\vec{\sigma})v \\ v \end{pmatrix} e^{i(Et-\vec{p}\vec{x})}$$

$$E = \sqrt{\vec{p}^2 + m^2}, \quad \vec{\eta} \equiv \frac{\vec{p}}{E+m}. \tag{A.15}$$

Dirac equation with electromagnetic interaction (Lorentz gauge ($\partial_\mu A^\mu = 0$)).

$$\begin{cases} \left(i\gamma_\mu\partial^\mu - m\right)\psi = e\gamma_\nu A^\nu \psi \\ \partial_\mu\partial^\mu A^\nu = e\{\overline{\psi}\gamma^\nu\psi\} \end{cases} \tag{A.16}$$

A.6   Two Spin Interaction

General formula.

$$\left\{w_i^\dagger\vec{\sigma}w_j\right\} \cdot \left\{w_k^\dagger\vec{\sigma}w_l\right\} \tag{A.17}$$

See Table A.1.

Spin magnetic moment-magnetic moment interaction.

$$(\vec{\sigma}_1 \cdot \vec{\sigma}_2) = \begin{pmatrix} 1 & 0 & 0 & 0 \\ 0 & -1 & 2 & 0 \\ 0 & 2 & -1 & 0 \\ 0 & 0 & 0 & 1 \end{pmatrix} \tag{A.18}$$

Table A.1 Elements of $\left\{ w_i^\dagger \vec{\sigma} w_j \right\} \cdot \left\{ w_k^\dagger \vec{\sigma} w_l \right\}$

	$\left\{ w_i^\dagger \vec{\sigma} w_j \right\}$	$\left\{ \chi_1^\dagger \vec{\sigma} \chi_1 \right\}$	$\left\{ \chi_1^\dagger \vec{\sigma} \chi_2 \right\}$	$\left\{ \chi_2^\dagger \vec{\sigma} \chi_1 \right\}$	$\left\{ \chi_2^\dagger \vec{\sigma} \chi_2 \right\}$
$\left\{ w_k^\dagger \vec{\sigma} w_l \right\}$		$\vec{e}_z$	$\vec{e}_-$	$\vec{e}_+$	$-\vec{e}_z$
$\left\{ \chi_1^\dagger \vec{\sigma} \chi_1 \right\}$	$\vec{e}_z$	1	0	0	-1
$\left\{ \chi_1^\dagger \vec{\sigma} \chi_2 \right\}$	$\vec{e}_-$	0	0	2	0
$\left\{ \chi_2^\dagger \vec{\sigma} \chi_1 \right\}$	$\vec{e}_+$	0	2	0	0
$\left\{ \chi_2^\dagger \vec{\sigma} \chi_2 \right\}$	$-\vec{e}_z$	-1	0	0	1

Appendix B

B

Abstract

This chapter describes how to solve the generalized Pauli equation.

B.1   Summary of Two Body Oscillation Formula

Before going to the detailed calculations, the results of the calculations are summarized here.

We deal with a system with two basis states, $|1\rangle$ and $|2\rangle$. The wave function of general state is expressed by a superposition of the basis vectors,

$$|\Psi[t]\rangle = c_1[t]|1\rangle + c_2[t]|2\rangle . \tag{B.1}$$

The coefficients satisfy the following generalized Pauli equation,

$$i\frac{d}{dt}\begin{pmatrix} c_1[t] \\ c_2[t] \end{pmatrix} = \begin{pmatrix} \mu_1 & \tau e^{-i\phi} \\ \tau e^{i\phi} & \mu_2 \end{pmatrix}\begin{pmatrix} c_1[t] \\ c_2[t] \end{pmatrix} . \tag{B.2}$$

The general wave function that satisfy the generalized Pauli equation (B.2) is

$$|\Psi[t]\rangle = \begin{pmatrix} \left(\cos\frac{\theta}{2}c_1[0] + e^{-i\phi}\sin\frac{\theta}{2}c_2[0]\right)\left(\cos\frac{\theta}{2}|1\rangle + e^{i\phi}\sin\frac{\theta}{2}|2\rangle\right)e^{-i(\mu_+ +\omega)t} \\ + \left(e^{i\phi}\sin\frac{\theta}{2}c_1[0] - \cos\frac{\theta}{2}c_2[0]\right)\left(e^{-i\phi}\sin\frac{\theta}{2}|1\rangle - \cos\frac{\theta}{2}|2\rangle\right)e^{-i(\mu_+ -\omega)t} \end{pmatrix}, \tag{B.3}$$

where

$$\mu_+ = \frac{\mu_1 + \mu_2}{2}, \quad \omega = \sqrt{\left(\frac{\mu_1 - \mu_2}{2}\right)^2 + \tau^2}, \quad \tan\theta = \frac{2\tau}{\mu_1 - \mu_2} . \tag{B.4}$$

The wave functions of the mass eigenstate are

$$\begin{cases} |\Psi_+[t]\rangle = (\cos[\theta/2]|1\rangle + e^{i\phi}\sin[\theta/2]|2\rangle)e^{-i((\mu_+ +\omega)t - \delta_+)} \equiv |+\rangle e^{-i(m_+ t - \delta_+)} \\ |\Psi_-[t]\rangle = (-e^{-i\phi}\sin[\theta/2]|1\rangle + \cos[\theta/2]|2\rangle)e^{-i((\mu_+ -\omega)t - \delta_-)} \equiv |-\rangle e^{-i(m_- t - \delta_-)} . \end{cases} \tag{B.5}$$

© Springer Nature Switzerland AG 2021
F. Suekane, *Quantum Oscillations*, Lecture Notes in Physics 985,
https://doi.org/10.1007/978-3-030-70527-5_B

The oscillation probability is

$$P_{|1\rangle \to |2\rangle}[t] = |\langle 2|\Psi[t]\rangle|^2 = \sin^2 \theta \sin^2 \omega t . \tag{B.6}$$

B.2   How to Solve the Generalized Pauli Equation

Generally the wave function of a system which is made of two basis states ($|1\rangle$, $|2\rangle$) is expressed as

$$|\Psi[t]\rangle = c_1[t]|1\rangle + c_2[t]|2\rangle , \tag{B.7}$$

where $c_i[t]$ are the amplitude for the system to be the state $|i\rangle$. The normalization condition is

$$|\Psi[t]|^2 = |c_1[t]|^2 + |c_2[t]|^2 = 1 \tag{B.8}$$

for any t. The generalized Pauli equation controls the time development of c_i,

$$i\frac{d}{dt}\begin{pmatrix} c_1[t] \\ c_2[t] \end{pmatrix} = \begin{pmatrix} \mu_1 & \tau e^{-i\phi} \\ \tau e^{i\phi} & \mu_2 \end{pmatrix}\begin{pmatrix} c_1[t] \\ c_2[t] \end{pmatrix} , \tag{B.9}$$

where μ_i, τ and ϕ are real numbers and the parameter domains are defined as ($\tau \geq 0$) and ($0 \leq \phi < 2\pi$). For the Pauli equation, $\mu_2 = -\mu_1$. The "generalized" means $-\mu_1$ and μ_2 can be different and the matrix in the equation is the most general 2×2 Hermitian matrix.

As a preparation to solve this equation, we rewrite the Eq. (B.9) as

$$
\begin{aligned}
i\begin{pmatrix} \dot{c}_1 \\ \dot{c}_2 \end{pmatrix} &= \left\{ \frac{\mu_1 + \mu_2}{2}I + \sqrt{\left(\frac{\mu_2 - \mu_1}{2}\right)^2 + \tau^2}\begin{pmatrix} (\mu_1 - \mu_2)/(2\sqrt{\ }) & (\tau/\sqrt{\ })e^{-i\phi} \\ (\tau/\sqrt{\ })e^{i\phi} & (\mu_2 - \mu_1)/(2\sqrt{\ }) \end{pmatrix} \right\}\begin{pmatrix} c_1 \\ c_2 \end{pmatrix} \\
&= \left\{ \mu_+ I + \omega \begin{pmatrix} \cos\theta & e^{-i\phi}\sin\theta \\ e^{i\phi}\sin\theta & -\cos\theta \end{pmatrix} \right\}\begin{pmatrix} c_1 \\ c_2 \end{pmatrix} ,
\end{aligned}
\tag{B.10}
$$

where $\mu_\pm$, ω and θ are defined as

$$\mu_\pm \equiv \frac{\mu_1 \pm \mu_2}{2}, \qquad \omega \equiv \sqrt{\mu_-^2 + \tau^2}, \qquad \tan\theta \equiv \frac{\tau}{\mu_-} . \tag{B.11}$$

Relations between θ, μ_-, τ and ω are pictorially shown by the mixing triangle in Fig. B.1. Next, we define the parameter d_i as follows,

$$d_i[t] \equiv e^{i\mu_+ t}c_i[t] . \tag{B.12}$$

By putting d_i in the Eq. (B.10), we obtain a little bit simpler Pauli equation for d_i,

$$\frac{d}{dt}\begin{pmatrix} d_1 \\ d_2 \end{pmatrix} = -i\omega \begin{pmatrix} \cos\theta & e^{-i\phi}\sin\theta \\ e^{i\phi}\sin\theta & -\cos\theta \end{pmatrix}\begin{pmatrix} d_1 \\ d_2 \end{pmatrix} . \tag{B.13}$$

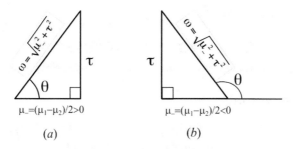

Fig. B.1 Mixing triangle. $\tan \theta \equiv \frac{\tau}{\mu_-}$, $\omega \equiv \sqrt{\mu_-^2 + \tau^2}$. **a** For $(\mu_1 > \mu_2)$ case, $(0 \leq \theta < \pi/2)$.**b** For $(\mu_1 < \mu_2)$ case, $(\pi/2 < \theta \leq \pi)$. For $(\mu_1 = \mu_2)$ case, $(\theta = \pi/2)$

In order to solve Eq. (B.13), we construct the following relation of d_i,

$$\frac{d}{dt}(d_1 + \lambda d_2) = -i\omega(\cos\theta + \lambda e^{i\phi}\sin\theta)\left(d_1 + \left(\frac{e^{-i\phi}\sin\theta - \lambda\cos\theta}{\cos\theta + \lambda e^{i\phi}\sin\theta}\right)d_2\right),$$

(B.14)

where λ is an arbitrary constant for now. If λ satisfies the following condition,

$$\lambda = \frac{e^{-i\phi}\sin\theta - \lambda\cos\theta}{\cos\theta + \lambda e^{i\phi}\sin\theta},$$

(B.15)

Equation (B.14) becomes

$$\frac{d}{dt}(d_1 + \lambda d_2) = -i\omega(\cos\theta + \lambda e^{i\phi}\sin\theta)(d_1 + \lambda d_2)$$

(B.16)

and the solution is

$$d_1[t] + \lambda d_2[t] = G\exp[-i\omega(\cos\theta + \lambda e^{i\phi}\sin\theta)t],$$

(B.17)

where G is an integral constant to be determined from a boundary condition.

Equation (B.15) can be rewritten as

$$\lambda^2 + 2\lambda e^{-i\phi}\cot\theta - e^{-2i\phi} = 0,$$

(B.18)

that has two solutions;

$$\begin{cases} \lambda_+ = e^{-i\phi}\tan\frac{\theta}{2} \\ \lambda_- = -e^{-i\phi}\cot\frac{\theta}{2}. \end{cases}$$

(B.19)

It is possible to show

$$\cos\theta + \lambda_\pm e^{i\phi}\sin\theta = \pm 1$$

(B.20)

and the corresponding solutions (B.17) are

$$\begin{cases} d_1 + e^{-i\phi}\tan\frac{\theta}{2}d_2 = G_+ e^{-i\omega t} \\ d_1 - e^{-i\phi}\cot\frac{\theta}{2}d_2 = G_- e^{+i\omega t}. \end{cases}$$

(B.21)

From this, the solution for d_i is

$$\begin{cases} d_1[t] = \cos^2 \frac{\theta}{2} G_+ e^{-i\omega t} + \sin^2 \frac{\theta}{2} G_- e^{i\omega t} \\ d_2[t] = e^{i\phi} \sin \frac{\theta}{2} \cos \frac{\theta}{2} (G_+ e^{-i\omega t} - G_- e^{i\omega t}) . \end{cases} \tag{B.22}$$

Therefore, from Eq. (B.12),

$$\begin{pmatrix} c_1[t] \\ c_2[t] \end{pmatrix} = e^{-i\mu_+ t} \begin{pmatrix} e^{-i\omega t} \cos^2 \frac{\theta}{2} & e^{i\omega t} \sin^2 \frac{\theta}{2} \\ e^{-i\omega t} e^{i\phi} \sin \frac{\theta}{2} \cos \frac{\theta}{2} & -e^{i\omega t} e^{i\phi} \sin \frac{\theta}{2} \cos \frac{\theta}{2} \end{pmatrix} \begin{pmatrix} G_+ \\ G_- \end{pmatrix} . \tag{B.23}$$

This is the general solution for the generalized Pauli equation (B.9).

In order to express the solution using the initial condition $c_i[0]$, we express $c_i[0]$ by $G_\pm$,

$$\begin{pmatrix} c_1[0] \\ c_2[0] \end{pmatrix} = \begin{pmatrix} \cos^2 \frac{\theta}{2} & \sin^2 \frac{\theta}{2} \\ e^{i\phi} \sin \frac{\theta}{2} \cos \frac{\theta}{2} & -e^{i\phi} \sin \frac{\theta}{2} \cos \frac{\theta}{2} \end{pmatrix} \begin{pmatrix} G_+ \\ G_- \end{pmatrix} . \tag{B.24}$$

From this,

$$\begin{pmatrix} G_+ \\ G_- \end{pmatrix} = \begin{pmatrix} 1 & e^{-i\phi} \tan \frac{\theta}{2} \\ 1 & -e^{-i\phi} \cot \frac{\theta}{2} \end{pmatrix} \begin{pmatrix} c_1[0] \\ c_2[0] \end{pmatrix} . \tag{B.25}$$

Therefore, by connecting Eqs. (B.23) and (B.25),

$$\begin{pmatrix} c_1[t] \\ c_2[t] \end{pmatrix} = e^{-i\mu_+ t} \begin{pmatrix} e^{-i\omega t} \cos^2 \frac{\theta}{2} + e^{i\omega t} \sin^2 \frac{\theta}{2} & (e^{-i\omega t} - e^{i\omega t}) e^{-i\phi} \sin \frac{\theta}{2} \cos \frac{\theta}{2} \\ (e^{-i\omega t} - e^{i\omega t}) e^{i\phi} \sin \frac{\theta}{2} \cos \frac{\theta}{2} & e^{-i\omega t} \sin^2 \frac{\theta}{2} + e^{i\omega t} \cos^2 \frac{\theta}{2} \end{pmatrix} \begin{pmatrix} c_1[0] \\ c_2[0] \end{pmatrix}$$

$$= \cdots = e^{-i\mu_+ t} \left(\cos \omega t I - i \sin \omega t \begin{pmatrix} \cos \theta & e^{-i\phi} \sin \theta \\ e^{i\phi} \sin \theta & -\cos \theta \end{pmatrix} \right) \begin{pmatrix} c_1[0] \\ c_2[0] \end{pmatrix} . \tag{B.26}$$

This is the same solution as (B.23) but a different expression.

B.3 Wave Functions That Satisfy the Pauli Equation

The general wave function can be expressed by several ways. Using (B.26),

$$|\Psi[t]\rangle = c_1[t]|1\rangle + c_2[t]|2\rangle = \cdots$$

$$= e^{-i\mu_+ t} \begin{pmatrix} \left((\cos \omega t - i \cos \theta \sin \omega t) c_1[0] - i e^{-i\phi} \sin \theta \sin \omega t c_2[0] \right) |1\rangle \\ + \left(-i e^{i\phi} \sin \theta \sin \omega t c_1[0] + (\cos \omega t + i \cos \theta \sin \omega t) c_2[0] \right) |2\rangle \end{pmatrix}$$

$$= \begin{pmatrix} \left(\cos \frac{\theta}{2} c_1[0] + e^{-i\phi} \sin \frac{\theta}{2} c_2[0] \right) \left(\cos \frac{\theta}{2} |1\rangle + e^{i\phi} \sin \frac{\theta}{2} |2\rangle \right) e^{-i(\mu_+ + \omega)t} \\ + \left(e^{i\phi} \sin \frac{\theta}{2} c_1[0] - \cos \frac{\theta}{2} c_2[0] \right) \left(e^{-i\phi} \sin \frac{\theta}{2} |1\rangle - \cos \frac{\theta}{2} |2\rangle \right) e^{-i(\mu_+ - \omega)t} \end{pmatrix} . \tag{B.27}$$

B.3.1 Energy Eigenstate

Wave function of the energy eigenstate with energy E has the general form,

$$\psi[t] = \text{(time independent term)} \times e^{-iEt}. \tag{B.28}$$

From (B.27), the wave function of the energy eigenstate with energy $E = \mu_+ + \omega$ is obtained under the condition that the term which is proportional to $e^{-i(\mu_+ - \omega)t}$ is zero,

$$e^{i\phi} \sin \tfrac{\theta}{2} c_1[0] - \cos \tfrac{\theta}{2} c_2[0] = 0. \tag{B.29}$$

From this, a relation between $c_1[0]$ and $c_2[0]$ is obtained as

$$c_2[0] = e^{i\phi} \tan \tfrac{\theta}{2} c_1[0]. \tag{B.30}$$

Combining with the normalization condition (B.8),

$$\left(1 + \tan^2 \tfrac{\theta}{2}\right) |c_1[0]|^2 = 1. \tag{B.31}$$

Then,

$$c_1[0] = e^{i\delta_+} \cos \tfrac{\theta}{2}, \quad c_2[0] = e^{i(\delta_+ + \phi)} \sin \tfrac{\theta}{2} \tag{B.32}$$

are obtained, where δ_+ is an arbitrary real number. Similarly, for $E = \mu_+ - \omega$ case, from

$$\cos \tfrac{\theta}{2} c_1[0] + e^{-i\phi} \sin \tfrac{\theta}{2} c_2[0] = 0, \tag{B.33}$$

the following coefficients are obtained,

$$c_1[0] = -e^{i(\delta_- - \phi)} \sin \tfrac{\theta}{2}, \quad c_2[0] = e^{i\delta_-} \cos \tfrac{\theta}{2}. \tag{B.34}$$

The wave functions of the energy eigenstate are then,

$$\begin{cases} |\Psi_+[t]\rangle = \left(\cos \tfrac{\theta}{2} |1\rangle + e^{i\phi} \sin \tfrac{\theta}{2} |2\rangle\right) e^{-i(m_+ t - \delta_+)} \\ |\Psi_-[t]\rangle = \left(-e^{-i\phi} \sin \tfrac{\theta}{2} |1\rangle + \cos \tfrac{\theta}{2} |2\rangle\right) e^{-i(m_- t - \delta_-)}, \end{cases} \tag{B.35}$$

where

$$m_\pm = \mu_+ \pm \omega. \tag{B.36}$$

Because $\mu_+ + \omega \geq \mu_+ - \omega$, the energy of Ψ_+ is, by definition, larger than that of Ψ_-.

The phases δ_+ and δ_- are independent and have to be determined by peripheral condition. The base vectors of the energy eigenstates are

$$\begin{cases} |+\rangle = e^{i\delta_+} \left(\cos \tfrac{\theta}{2} |1\rangle + e^{i\phi} \sin \tfrac{\theta}{2} |2\rangle\right) \\ |-\rangle = e^{i\delta_-} \left(-e^{-i\phi} \sin \tfrac{\theta}{2} |1\rangle + \cos \tfrac{\theta}{2} |2\rangle\right). \end{cases} \tag{B.37}$$

The expression (B.37) can generally be used at any situation. It can also be expressed by matrix form;

$$
\begin{pmatrix} e^{-i\delta_+} |+\rangle \\ e^{-i\delta_-} |-\rangle \end{pmatrix} = \begin{pmatrix} \cos\frac{\theta}{2} & e^{i\phi}\sin\frac{\theta}{2} \\ -e^{-i\phi}\sin\frac{\theta}{2} & \cos\frac{\theta}{2} \end{pmatrix} \begin{pmatrix} |1\rangle \\ |2\rangle \end{pmatrix}.
\tag{B.38}
$$

The matrix is called "mixing matrix".

A particular wave function can be expressed by both the energy eigenstates $|\pm\rangle$ and the original states $|1\rangle$, $|2\rangle$ as

$$
|\Psi[t]\rangle = c_1[t]|1\rangle + c_2[t]|2\rangle = c_+[0]e^{-im_+t}|+\rangle + c_-[0]e^{-im_-t}|-\rangle
\tag{B.39}
$$

The relation between $c_\pm$ and c_1, c_2 can be obtained from

$$
\begin{aligned}
c_1[t] &= \langle 1|\Psi[t]\rangle = c_+[0]e^{-im_+t}\langle 1|+\rangle + c_-[0]e^{-im_-t}\langle 1|-\rangle \\
&= \cos\tfrac{\theta}{2}c_+[0]e^{-i(m_++t-\delta_+)} - e^{-i\phi}\sin\tfrac{\theta}{2}c_-[0]e^{-i(m_--t-\delta_-)} \\
c_2[t] &= \langle 2|\Psi[t]\rangle = c_+[0]e^{-im_+t}\langle 2|+\rangle + c_-[0]e^{-im_-t}\langle 2|-\rangle \\
&= e^{i\phi}\sin\tfrac{\theta}{2}c_+[0]e^{-i(m_++t-\delta_+)} + \cos\tfrac{\theta}{2}c_-[0]e^{-i(m_--t-\delta_-)} .
\end{aligned}
\tag{B.40}
$$

The relation of the coefficients can be written simpler as

$$
\begin{pmatrix} c_1[t] \\ c_2[t] \end{pmatrix} = \begin{pmatrix} \cos\frac{\theta}{2} & -e^{-i\phi}\sin\frac{\theta}{2} \\ e^{i\phi}\sin\frac{\theta}{2} & \cos\frac{\theta}{2} \end{pmatrix} \begin{pmatrix} c_+[0]e^{-i(m_++t-\delta_+)} \\ c_-[0]e^{-i(m_--t-\delta_-)} \end{pmatrix}.
\tag{B.41}
$$

This form indicates that $\delta_\pm$ are the phases of the energy eigenstate at $t = 0$. This matrix is also called the mixing matrix. The inverse relation of (B.38) is

$$
\begin{pmatrix} |1\rangle \\ |2\rangle \end{pmatrix} = \begin{pmatrix} \cos\frac{\theta}{2} & -e^{i\phi}\sin\frac{\theta}{2} \\ e^{-i\phi}\sin\frac{\theta}{2} & \cos\frac{\theta}{2} \end{pmatrix} \begin{pmatrix} e^{-i\delta_+}|+\rangle \\ e^{-i\delta_-}|-\rangle \end{pmatrix}
\tag{B.42}
$$

and this mixing matrix is similar to that in Eq. (B.41) and when dealing with the mixing matrix, we have to be careful which one (mixing matrix for base vectors or that for amplitudes) is being used.

B.3.2 Oscillation

If the state is $|1\rangle$ at time $t = 0$, the initial condition is expressed as $c_1[0] = 1$ and $c_2[0] = 0$, and the wave function (B.27) at time t becomes

$$
|\Psi[t]\rangle = e^{-i\mu+t}((\cos\omega t - i\sin\omega t\cos\theta)|1\rangle - i\sin\omega t e^{i\phi}\sin\theta|2\rangle) .
\tag{B.43}
$$

The probability that the state changes to $|2\rangle$ after time t is then,

$$
P_{|1\rangle\to|2\rangle}[t] = |\langle 2|\Psi[t]\rangle|^2 = \sin^2\theta\sin^2\omega t .
\tag{B.44}
$$

The probability oscillates with angular velocity 2ω and probability amplitude $\sin^2\theta$.

Appendix C

C

C.1 Summary of Oscillations and Mixings

© Springer Nature Switzerland AG 2021
F. Suekane, *Quantum Oscillations*, Lecture Notes in Physics 985,
https://doi.org/10.1007/978-3-030-70527-5

Table C.1 Transitions, energy eigenstates and their origins

Name	Origin	Transition	Energy eigenstate
Neutrino oscillation	X	$|\nu_e\rangle \leftrightarrow |\nu_\mu\rangle \leftrightarrow |\nu_\tau\rangle$	ν_1, ν_2, ν_3
Cabbibo angle	Higgs	$|d'\rangle \leftrightarrow |s'\rangle$	d, s
Chirality Osc.	Higgs	$|L\rangle \leftrightarrow |R\rangle$	$|R\rangle \pm |L\rangle$
Weinberg angle	Higgs	$W_3 \leftrightarrow B$	γ, Z^0
Hydrogen 21 cm line	$\vec{\mu}_p \cdot \vec{\mu}_e$	$|p(\Uparrow)e(\Downarrow)\rangle \leftrightarrow |p(\Downarrow)e(\Uparrow)\rangle$	$|\Uparrow\Downarrow\rangle \pm |\Downarrow\Uparrow\rangle$
$\pi^+ - \rho^+$ mass difference	Strong	$|\Uparrow\Downarrow\rangle_S \leftrightarrow |\Downarrow\Uparrow\rangle_S$	π^+, ρ^+
CPV	Weak	$K_+^{CP} \leftrightarrow K_-^{CP}$	K_1, K_2
Positronium	EM	$|e^+e^-\rangle \leftrightarrow |e^-e^+\rangle$	o-Ps, p-Ps
Baryon color	Strong	$|RGB\rangle \leftrightarrow |GRB\rangle$	$|RGB\rangle - |RBG\rangle + |BRG\rangle - \cdots$
ρ^0, ω, ϕ structure	Strong	$|u\bar{u}\rangle \leftrightarrow |d\bar{d}\rangle \leftrightarrow |s\bar{s}\rangle$	ρ^0, ω, ϕ
Spin precession in $\vec{B}$	$\vec{\mu} \cdot \vec{B}$	$|\Uparrow\rangle \leftrightarrow |\Downarrow\rangle$	$\hat{s}[\theta, \phi]$
$g - 2$	Higher order	$|\Uparrow\rangle \leftrightarrow |\Downarrow\rangle$	

Index

A
AGS, 109
Amplitude, 1
Anomalous magnetic moment
 lepton, 39, 42, 43
 proton (κ_p), 28, 50

B
Baryon, 131
Basis state, 166
β decay, 88
Bohr magneton (μ_B), 9, 10, 27
Bohr radius (a_B), 29

C
Cabibbo angle (θ_C), 79, 82, 84, 97, 171
Cabibbo–Kobayashi–Maskawa matrix, 79, 87
Charm quark, 102
Chirality, 66, 73
 left-handed, 67, 73
 right-handed, 67, 74
 swap, 75
 transition, 74
CKM matrix, 87, 104, 110
CKM mixing, 79
Color, 171
Constituent quark mass, 113, 114
Coupling
 quark-pion, 134
 quark-$W^{\pm}$, 102
Coupling constant
 strong, 115

Covariant derivative, 58
CP eigenstate, 107
CP violation, 92, 97, 104, 108, 109, 158, 171
 direct, 110
 indirect, 108
CPLEAR, 100
CPT invariance, 155
CPT theorem, 99
Cross transition, 4, 11, 63, 146
Current quark mass, 113
Cyclotron frequency, 39

D
Dark matter, 37
Decay constant, 102
Dirac equation, 9, 10, 58, 73–75, 163
Dirac fermion, 39
Dirac matrix, 74, 162
Direct neutrino mass measurement, 152
Doppler shift, 38
Down-type quark, 79

E
e^+e^- annihilation, 47, 73
Effective mass
 neutrino, 152
Electromagnetic interaction, 57, 171
Electrostatic potential, 49
Electroweak parameter, 67
Energy eigenstate, 3, 12, 15, 18, 21, 169

© Springer Nature Switzerland AG 2021
F. Suekane, *Quantum Oscillations*, Lecture Notes in Physics 985,
https://doi.org/10.1007/978-3-030-70527-5

F
Fermi constant (G_F), 69, 102
Flavor, 145
Flavor mass of neutrino, 151
Forward-Backward asymmetry (A_{FB}), 69

G
$g - 2$, 43
    electron, 45
    muon, 43
Gauge boson, 57
g factor, 42
Gluon, 115

H
Helicity, 40, 43
    conservation, 39, 40
HFS, 50
Higgs boson, 57, 63, 71
Higgs field, 2, 76, 80, 171
Higgs vacuum expectation value (v_0), 64, 80, 84
HI line, 27, 37, 117
Hydrogen
    21cm line, 37, 171
    magnetic moment, 33
    spin structure, 27
Hyperfine splitting, 50, 51

I
Imaginary mass, 76
Interaction
    magnetic moment-magnetic moment, 27, 117, 139
Isospin, 139
Isosymmetry, 115, 120

K
Klein-Gordon equation, 64

L
Left-Right asymmetry (A_{LR}), 68
LHC, 73, 76
Lorentz factor, 149
Lorentz transformation, 149

M
Magnetic moment, 10
Mass difference

π^+-ρ^+, 171
Mass eigenstate, 3, 12, 35
    K_1, K_2, 171
    γ, Z^0, 171
    $\hat{s}[\theta, \phi]$, 171
    $|L\rangle \leftrightarrow |R\rangle$, 171
    $|RGB\rangle - |RBG\rangle + |BRG\rangle - \cdots$, 171
    $|\Uparrow\Downarrow\rangle \pm |\Downarrow\Uparrow\rangle$, 171
    ν_1, ν_2, ν_3, 171
    π^+, ρ^+, 171
    ρ^0, ω, ϕ, 171
    d, s, 171
    o-Ps, p-Ps , 171
Meson, 113
Mixing matrix, 4, 170
Mixing triangle, 25, 35, 60, 166
    (ν_e, ν_μ), 146
    relativistic 2 flavor neutrinos, 151
MM, 27
MM-MM interaction, 27, 28, 47, 51, 117, 139
Muon decay, 76
Muon polarization, 44

N
Neutral current, 67
Neutrino, 145
Normalization condition, 166
Nuclear magneton (μ_N), 27

O
Ortho-positronium (o-Ps), 53
Oscillation, 16, 22, 32, 66, 85, 170
    3 flavor neutrinos, 152
    quark, 93
    $B^0 \leftrightarrow \overline{B^0}$, 91
    chirality, 171
    $K^0 \leftrightarrow \overline{K^0}$, 97, 99
    $K_+^{CP} \leftrightarrow K_-^{CP}$, 108
    neutrino, 145, 171
    $\nu_e \leftrightarrow \nu_\mu$, 145
    relativistic, 148
    6 quark system, 106
    two body formula, 165
    $|u\overline{u}\rangle \leftrightarrow |d\overline{d}\rangle$, 121
Oscillation period, 37
Oscillation probability
    $\nu_\mu \rightarrow \mu_e$, 147, 148
    relativistic 2 flavor neutrinos, 149

P
Pair annihilation, 51

Pair creation, 47
Para-positronium (p-Ps), 53
Pauli equation, 9, 10, 14, 19, 140, 166
    generalized, 165, 166
Pauli exchange principle, 108, 131
Pauli matrix, 30, 59, 162
Pauli spin vector, 48
Physics parameters, 161
π meson exchange, 134
PMNS matrix, 154
Pontercorvo-Maki-Nakagawa-Sakata (PMNS)
    matrix, 154
Positronium (Ps), 47, 49, 171
Potential
    electromagnetic, 113
    strong, 113
Pseudoscalar meson, 113, 119

Q
QCD, 113
QED, 29, 47, 59
Quantize, 9, 10, 29, 33
Quantum oscillation, 1, 5
Quark
    d, s, b, u, c, t, 79

R
Radiative correction, 71
Reduced mass, 49
Relativistic correction, 53
Relativistic quantum mechanics, 11
RHC, 74, 76
ρ^+, 116
Right-handed, 73
Running α_S, 117

S
S-channel, 47
Schrödinger equation, 1, 49
Schrödinger-Pauli equation, 40
Self coupling of gluons, 117
Self transition, 4, 11, 52, 63, 75, 98
    $|G\overline{G}\rangle$, 128
    $|RG\rangle$, 132
    $|sdd\rangle$, 137
    $|sud\rangle$, 137
    $|suu\rangle$, 137
    $|ud\rangle$, 134
    $|uu\rangle$, 135
    $(|u\overline{u}\rangle \pm |d\overline{d}\rangle)/\sqrt{2}$, 123
    (u, c, t), 86
    $|u\overline{d}\rangle$, 115

Six quark system, 86
Spherical harmonics, 49
Spin
    bssis vector, 15
    direction, 12, 22
    flip, 2, 11, 16
    precession, 13, 22, 39, 171
    transition, 11
    transition amplitude, 14, 30
    wave function, 13, 15, 19, 22
Spin direction vector $(\vec{e}_s)$, 162
Spinor, 161
    four component-, 10
    two component-, 10
Spin-0 state, 50
Spin-1 state, 50
Standard model, 57, 79, 145
State mixing, 1
Storage ring, 44
Strong
    electrostatic field, 117
    magnetic moment, 117
    potential, 115
Strong coupling constant (α_S), 117
Strong interaction, 2, 171
SU(6), 137
SU(2)$_{\text{spin}}$, 137
SU(3)$_{\text{flavor}}$, 137
Superposition, 1, 16, 47, 63
    $(|\nu_e\rangle, |\nu_\mu\rangle))$, 146
    $(|R\rangle, |L\rangle)$, 74
    (d, s), 102
    $(K^0, \overline{K^0})$, 97
    $(K_+^{\text{CP}}, K_-^{\text{CP}})$, 110

T
T-channel, 47
Time delation, 40
Top quark, 57, 70, 104
Totally antisymmetric, 131
Transition
    $K_+^{\text{CP}} \leftrightarrow K_-^{\text{CP}}$, 171
    $W_3 \leftrightarrow B$, 171
    $|L\rangle \leftrightarrow |R\rangle$, 171
    $|RGB\rangle \leftrightarrow |GRB\rangle$, 171
    $|\nu_e\rangle \leftrightarrow |\nu_\mu\rangle \leftrightarrow |\nu_\tau\rangle$, 171
    $|\Uparrow\Downarrow\rangle_S \leftrightarrow |\Downarrow\Uparrow\rangle_S$, 171
    $|\Uparrow\rangle \leftrightarrow |\Downarrow\rangle$, 171
    $|e^+e^-\rangle \leftrightarrow |e^-e^+\rangle$, 171
    $|p(\Uparrow)e(\Downarrow)\rangle \leftrightarrow |p(\Downarrow)e(\Uparrow)\rangle$, 171
    $|u\overline{u}\rangle \leftrightarrow |d\overline{d}\rangle \leftrightarrow |s\overline{s}\rangle$, 171
    $(|u\overline{u}\rangle \pm |d\overline{d}\rangle)/\sqrt{2} \leftrightarrow |s\overline{s}\rangle$, 123
    chirality, 74

(ψ_L, ψ_R), 75
$\overline{K^0} \to K^0$, 105
antiquark, 140
$K^0 \leftrightarrow \overline{K^0}$, 97
$K^0 \to K^0$, 105
$|du\rangle \leftrightarrow |ud\rangle$, 137
$|G\overline{G}\rangle \leftrightarrow |R\overline{R}\rangle$, 128
$|+\rangle \leftrightarrow |s\overline{s}\rangle$, 124
$|RG\rangle \leftrightarrow |GR\rangle$, 132
$|su\rangle \leftrightarrow |us\rangle$, 137
$|ud\rangle \leftrightarrow |du\rangle$, 134
$|u\overline{u}\rangle \leftrightarrow |d\overline{d}\rangle$, 120
$|u\overline{u}\rangle \leftrightarrow |d\overline{d}\rangle \leftrightarrow |s\overline{s}\rangle$, 122, 126
$\nu_e \leftrightarrow \nu_\mu$, 146
quark, 139
spin, 41
Transition amplitude, 2, 19
    3 flavor neutrinos, 153, 159
    D-type quarks, 93
    U-type quarks, 93
Transition diagram, 3
    CP base, 107
    6 quark K^0-$\overline{K^0}$, 105
Transition equation, 2, 87, 92, 99
    2 flavor neutrinos, 153
    3 flavor neutrinos, 152
    CP base, 107
    $\nu_e \leftrightarrow \nu_\mu$, 146

6 quark K^0-$\overline{K^0}$, 105
Two-component spinor, 73
Two spin interaction, 163

U
Uncertainty principle, 37, 85
Up-type quark, 79

V
Vector meson, 113, 119
Vertex
    fermion-photon, 116
    QCD, 116
    QED, 40, 116
    quark-gluon, 116
Vertex correction, 42

W
W boson, 44
Weak eigenstate, 79
Weak interaction, 57, 73, 79, 171
Weak mixing angle (θ_W), 57
Weinberg angle (θ_W), 57, 61, 171

Y
Yukawa coupling, 76

Printed in the United States
by Baker & Taylor Publisher Services